Avian Ecology in Latin American Cityscapes

Ian MacGregor-Fors • Juan F. Escobar-Ibáñez
Editors

Avian Ecology in Latin American Cityscapes

Springer

Editors
Ian MacGregor-Fors
Red de Ambiente y Sustentabilidad
Instituto de Ecología, A.C.
Xalapa, Veracruz, Mexico

Juan F. Escobar-Ibáñez
Red de Ambiente y Sustentabilidad
Instituto de Ecología, A.C.
Xalapa, Veracruz, Mexico

ISBN 978-3-319-63473-9 ISBN 978-3-319-63475-3 (eBook)
DOI 10.1007/978-3-319-63475-3

Library of Congress Control Number: 2017950290

Printed on acid-free paper

This Springer imprint is published by Springer Nature
The registered company is Springer International Publishing AG
The registered company address is: Gewerbestrasse 11, 6330 Cham, Switzerland

To my muses, Nina, Alicia, and Gloria

Ian MacGregor-Fors

To my parents, Gabriela and Fernando

Juan F. Escobar-Ibáñez

Guacamaya, guacamaya, no vayas a la ciudad;
no vayas a la ciudad, guacamaya, guacamaya.
¿Qué buscas allá malhaya?, la vida te costará;
sólo por unas pitayas, que en ese monte no
hay ya.[1]

Son Jarocho

[1] Mexican regional song that picturesquely adverts macaws of urban threats, regardless of the presence of feeding resources in the city.

Foreword

Creating Cities Where the Pines Flourish and Jays Still Scream

Latin America is rapidly becoming urbanized. It includes two of the world's largest cities, Mexico City and São Paulo, as well as some 60 cities with populations over one million inhabitants and growing. The growth and expansion of cities globally is stressing existing social, economic, and ecological systems. Latin American cities currently face a number of formidable challenges related to increasing energy demands, lack of public transportation, increasing air and water pollution, poor waste management, poor food quality and quantity, and housing shortages. Separately or in combination, these challenges have a significant impact on the health and well-being of Latin American urbanites. The creation of this book was motivated, in part, by the fact that this region is home to some of the most biodiversity-rich ecosystems in the world, especially in regard to bird diversity. The ensuing land-use and land-cover changes, alterations to biogeochemical cycles, and biological invasions resulting from rapid urbanization of the region also have the potential to significantly reduce the local, regional, and global avian diversity in the future.

The loss of avian diversity in Latin America due to urbanization will result in cultural, economic, and ecological consequences that will likely produce negative local and global impacts. This is because birds have intrinsic value as components of complex natural ecosystems as well as ethical, moral, and aesthetic values. In addition, as described in the 2005 Millennium Ecosystem Assessment, they provide a variety of supporting, provisioning, regulating, and cultural ecosystem services. Recent research has provided growing evidence that maintaining biodiversity in cities, especially plants, birds, and mammals, increases the well-being of urban dwellers. This research supports E.O. Wilson's 1984 'biophilia hypothesis' that contends people to have an innate desire to have contact with nature. It is interesting to note that as early as 1849 Henry David Thoreau understood the value of maintaining plants and animals in human settlements:

"Even the oldest villages are indebted to the border of wild wood which surrounds them, more than to the gardens of men. There is something indescribably inspiriting and beautiful in the aspect of the forest skirting and occasionally jutting into the midst of new towns, which, like the sand-heaps of fresh fox-burrows, have sprung up in their midst. The very

uprightness of the pines and maples asserts the ancient rectitude and vigor of nature. Our lives need the relief of such a background, where the pine flourishes and the jay still screams."

Henry David Thoreau, 1849. *A Week on the Concord and Merrimack Rivers*

In order to create Latin American cities "where the pine flourishes and the jay still screams," we need to incorporate evidence-based urban ecological research into the design, construction, and management of cities in the future. Ian MacGregor-Fors and Juan F. Escobar-Ibáñez have assembled an impressive group of ornithologists and have compiled the first comprehensive assessment of the current state of knowledge of urban birds in Latin America. One of the earliest assessments of the ecology of urban birds by Marzluff, Bowman, and Donnelly published in 2001, entitled *Avian Ecology and Conservation in an Urbanizing World*, and the recently (2017) edited book by Murgui and Hedblom, entitled *Ecology and Conservation of Birds in Urban Environments*, provide global assessments of urban avian ecology research but include relatively little information from Latin American cities. The current volume effectively adds critical new information about the avian ecology of Latin American cities which will provide researchers the ability to evaluate whether research results and conservation strategies obtained from studies from cities in Northern America, Europe, and Australia apply to Latin America.

Ecological studies of Latin American cities were commenced only about 30 years ago but have been increasing in number over the last five years. Interestingly, the ecology of urban birds is one of the most common research subjects. The authors readily admit there are more studies of urban birds in North America and Europe, but there are enough existing Latin American bird studies to make appropriate global comparisons. All of the chapters in the book provide important information regarding the ecology of birds in Latin American cities including topics such as species richness, composition and abundance, demography and population dynamics, behavioral responses, the influence of vegetation cover and built surfaces on bird distribution and abundance, urban threats to birds, and the conservation of birds in urban environments. The editors have done an excellent job of synthesizing the key messages, and they have provided useful insights into the current gaps in knowledge and methodologies which will facilitate the identification of new research directions as well as management and conservation practices. In conclusion, the authors have produced an important compilation of information that greatly enhances our knowledge of the avian ecology of Latin American cities. This is a critical first step in developing the designs, tools, and conservation strategies for creating bird-friendly cities in the future.

Mark J. McDonnell, Editor-in-Chief,
Journal of Urban Ecology,
Oxford University Press,
Oxford, UK

Acknowledgments

Firstly, we would like to thank our colleagues that made this book possible. All contributors worked closely with us to develop an interconnected piece that weaves through urban bird studies in our beloved Latin America, *¡Infinitas gracias a todos! Muito obrigado a todos!*

We were very fortunate to have an international expert panel of peer reviewers – to whom we are deeply grateful – who provided open, detailed, and constructive comments and suggestions that undoubtedly enhanced the structure, quality, and clarity of all chapters (see list of reviewers below).

We are highly honored to have Mark J. McDonnell frame our book with such an eloquent Foreword. We hope that this book encourages not only ecologists in Latin America to explore urban systems but also colleagues from understudied regions where urbanization and economic inequality or poverty meet megadiversity.

We are also greatly thankful to the many people that were involved in figure making, proofreading, and reference checking, as well as providing insightful recommendations for the book, in particular, Rafael Rueda Hernández, Ina Falfán, Michelle García Arroyo, Jeffrey H. Lee, Julian Avila Campos, Miguel A. Gómez Martínez, Oscar H. Marín Gómez, and Carlos M. Trujillo Torres.

Finally, we would like to express our gratitude to Springer, who through João V. Pildervasser (Associate Editor, Life Sciences) encouraged us to develop a book focused on bird studies in urban Latin America, as well as to the Springer Production Team, particularly Susan Westendorf (Project Coordinator, Book Production) and Rajagopal Pavithra (Project Manager), who guided us though the editorial process.

Panel of International Expert Peer Reviewers
Karl L. Evans – *University of Sheffield, UK*
Mark A. Goddard – *Newcastle University, UK*
Amy K. Hahs – *Urban Ecology in Action, Melbourne, Australia*
Madhusudan Katti – *North Carolina State University, USA*
Charles H. Nilon – *University of Missouri-Columbia, USA*
Yolanda van Heezik – *University of Otago, New Zealand*
Paige S. Warren – *University of Massachusetts-Amherst, USA*
Rachel White – *Santa Fe, New Mexico, USA*

Contents

Chapter 1
Birds from Urban Latin America, Where Economic Inequality and Urbanization Meet Biodiversity

Ian MacGregor-Fors and Juan F. Escobar-Ibáñez

Abstract Following a massive migration toward cities, more than half of the total human population is now urban. Given the unbalanced metabolism of urban systems, urbanization has been recognized to be a major ecological driver with worrisome consequences at different spatial and temporal scales. Such an environmental issue has drawn the attention of ecologists from across the globe since the late 1990s, when urban ecology consolidated as a discipline. Although urban ecology has developed importantly in the last three decades, our comprehension of the effects of urbanization on wildlife communities is heavily biased toward developed countries, most of which are located in temperate regions. Woefully, there is a dearth of knowledge from highly biodiverse areas with increasing urbanization rates and major urban agglomerations. Latin America is not an exception, with most of its population in rapidly growing urban centers. Given that the region concentrates important biodiverse areas in a scenario of considerable economic and social disparity, understanding the effects that urbanization has on wildlife species is of special concern. Even though ecological studies performed in urban Latin America started in the 1970s, urban ecology in the region is still in the process of consolidation, with birds being the most studied group. Several ecological patterns hold in urban Latin America when contrasted with those from other regions; yet, important differences have been identified, making evident the need to understand the response of wildlife species in the region. In this book, we gathered regional experts to set the state-of-the-art of bird studies in urban Latin America. Starting with an updated review, the book transits across topics such as urban bird species richness, composition, abundance, demography, population dynamics, behavior, threats, and conservation, as well as their relationships with 'green' and 'gray' infrastructures. After reviewing the specific topics with information from across the globe, each chapter contrasts the global findings with those from Latin America, identifying knowledge gaps and research needs to suggest future directions. The gathered information sets the

I. MacGregor-Fors (✉) • J.F. Escobar-Ibáñez
Red de Ambiente y Sustentabilidad, Instituto de Ecología, A.C.
Carretera antigua a Coatepec 351, El Haya, 91070 Xalapa, Veracruz, Mexico
e-mail: ian.macgregor@inecol.mx

I. MacGregor-Fors, J.F. Escobar-Ibáñez (eds.), *Avian Ecology in Latin American Cityscapes*, DOI 10.1007/978-3-319-63475-3_1

foundation for the study of birds in urban Latin America, hopefully promoting the consolidation of the field in the region and encouraging future mechanistic studies that untangle the recorded patterns to have the required information to bridge the gap between evidence-based knowledge and practice in urban systems.

Keywords Anthropocene • Avian studies • Ecological footprint • Megadiverse cities • Urban ecology

1.1 Urbanization: Ecological Consequences and Study Paradigms

The human ecological footprint is currently ubiquitous, with increasing disturbance intensity and frequency resulting in unquantifiable effects at different spatial and temporal scales (Bazzaz 1990; Steffen et al. 2005; Maxwell et al. 2016; Venter et al. 2016). Human-related activities have modified ecosystems, directly or indirectly, to such an extent that there are no current ecosystems that can be considered pristine (Vitousek et al. 1997; Steffen et al. 2015). The consequences of human activities are of such magnitude that scientists have recognized the beginning of a new era triggered, among other causes, by the Industrial Revolution: The Anthropocene (Crutzen and Stoermer 2000; Steffen et al. 2011; Waters et al. 2016). Among the most impressive human phenomena of modern history, the wave of migration from nonurban areas (e.g., agricultural settlements, small towns) to urban centers is notable (United Nations 2014, 2015a). Such a massive migration, mainly driven by modern industrial and socioeconomic dynamics, has resulted in an important shift of the total human population, passing from being predominantly nonurban ('rural') to mainly urban (Grimm et al. 2008; United Nations 2015a). As a direct response of the human concentration in cities, their borders are sprawling, and new urban systems are establishing faster than ever, requiring disproportionate amounts of resources to fulfill the modern urban lifestyle (Grimm et al. 2008; Seto et al. 2011). Not surprisingly, the footprint of a city, related to the food, energy, and materials that are imported into the system, often exceeds 200 times its territory (Wigginton et al. 2016).

When an area is urbanized, drastic changes occur at different spatial and temporal scales, including the removal of preexistent vegetation, important shifts in the terrain, redirecting of rivers, filling of water bodies, as well as the addition of artificial structures that result in large extensions of hard, relatively impervious, surfaces (Eldredge and Horenstein 2014). Such drastic changes in the land, most of which are long-lasting, have been related to alarming ecological and human health issues (McKinney 2006; Tzoulas et al. 2007; Seto et al. 2011; McDonnell and MacGregor-Fors 2016). Although the global urban land only represented 3% of the terrestrial surface by 2010 (with only 0.45% covered by impervious surfaces; Liu et al. 2014), the ecological consequences of urbanization and its processes go way beyond their physical limits (Berkowitz et al. 2003). In fact, urbanization has been related to some of the most alarming components of global change (i.e., land use and land cover transformation, climate change, biological invasions, changes in biogeochemical cycles, biodiversity loss; Grimm et al. 2008) and also identified as one of the major causes of species endangerment (Czech et al. 2000; Marzluff et al. 2001; Seto et al. 2011; Maxwell et al. 2016).

All of the above has led to great concern in the scientific community, with ecologists consolidating urban ecology as a discipline after the urban development explosion of the 1990s (McDonnell 2011). Since the foundation of urban ecology, it has been well recognized that cities are comprised of three major spheres: physical, ecological, and social (Alberti et al. 2003; Berkowitz et al. 2003; Nilon et al. 2011). Since the 1990s, a vision of urban ecological studies as generators of information to have a direct impact on human well-being through the maintenance of ecological processes within cities was established (Niemelä 1999). Yet, most studies have been framed following two approaches: ecology 'in' and 'of' the city (Pickett et al. 1999). Briefly, the ecology 'in' the city paradigm focuses on biologically oriented studies performed in urban systems to assess ecological questions, while the ecology 'of' the city paradigm broadens the scope to include the complexity of urban systems, considering all three major spheres and their interactions (Pickett et al. 2011). Most recently, Pickett et al. (2016) highlighted an emerging paradigm, ecology 'for' the city, that encourages ecologists to collaborate with specialists from other disciplines, as well as all implied urban stakeholders to promote livable, sustainable, and healthier cities through environmental integrity, social equity, and economic viability.

1.2 Latin America, a Complex Scenario for Biodiversity

In Latin America, economic inequity, urbanization, and biodiversity converge (Santandreu et al. 2002). The region experiences important urban growth, housing more than 500 million urbanites (United Nations 2015b). A considerable proportion of the Latin American population is concentrated in urban centers (80%), four of which are among the 28 'megacities' (agglomerations with >10 million inhabitants) recognized by the United Nations (2014) (i.e., Mexico City, Mexico: 20.8 million; São Paulo, Brazil: 20.8 million; Buenos Aires, Argentina: 15.0 million; Rio de Janeiro, Brazil: 12.8 million), and another four among the 72 urban centers with >5 million inhabitants (i.e., Lima, Peru: 9.7 million; Bogota, Colombia: 9.5 million, Santiago, Chile: 6.5 million; Belo Horizonte, Brazil: 5.7 million). It is forecast that the population of these eight cities will have a ~15% average growth by 2030. Moreover, two Mexican cities are predicted to be included among the 100 most populated cities of the world by 2030: Guadalajara (5.8 million) and Monterrey (5.5 million; United Nations 2015b; Fig. 1.1).

Together with the overcrowding of cities and the rapid urbanization occurring in Latin America, the region is marked by extreme inequities (Pauchard and Barbosa

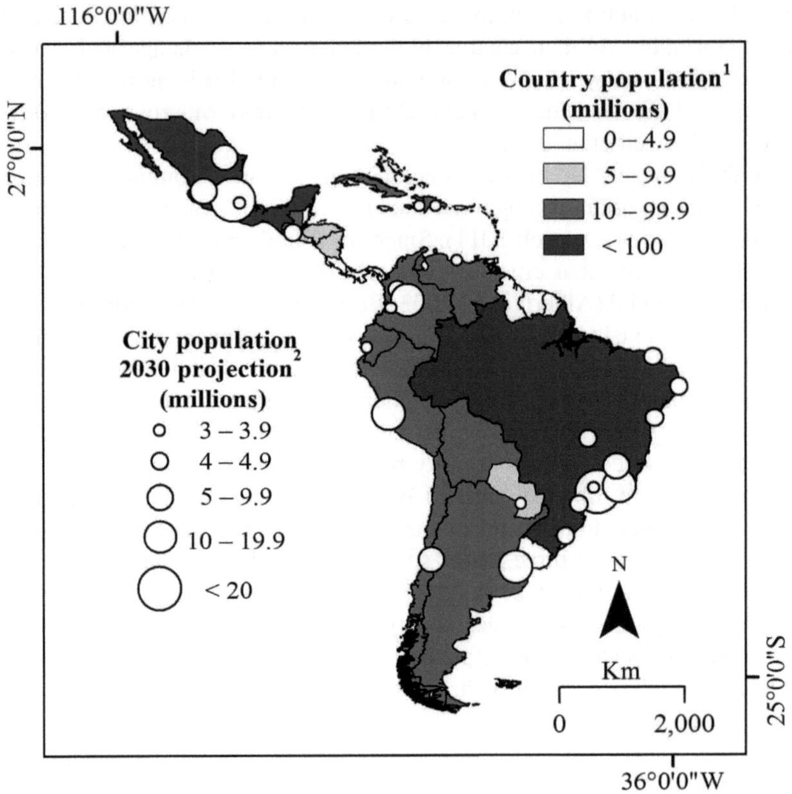

Fig. 1.1 Human population in urban Latin America highlighting [1]2014 values for countries and [2]2030 projections for cities with >3 million inhabitants

2013). Such disparities are of both social and economic nature, with more than 120 million urbanites living in poverty, largely as a result of employment deficits and economic informality. However, the urban scenario is not consistent throughout the region, with economic inequality being both higher and lower when compared to national calculations (e.g., higher Gini index values – measures of wealth distribution, for Guadalajara; Managua, Nicaragua; São Paulo; and lower values in Bogota, Santiago, Mexico City; ONU-Habitat 2012). Inequity is also translated to the ecological sphere of Latin American cities, with an uneven distribution of greenspaces, often concentrated in high-income areas (Barbosa et al. 2007; Reyes and Figueroa 2010).

This scenario of human agglomeration and social and economic inequities meets megadiversity in the region. Latin America, with seven out of 25 global biodiversity hotspots (Cincotta et al. 2000) and six out of 17 megadiverse countries (i.e., Brazil, Colombia, Ecuador, Mexico, Peru, and Venezuela; Mittermeier et al. 1997), holds an important proportion of the global terrestrial ecoregions (Olson and Dinerstein

2002), which are critical for global biodiversity conservation (Mittermeier et al. 2011). Regarding birds, one of the best studied wildlife groups due to their potential as bioindicators (Moreno et al. 2007), most of their diversity and endemicity are concentrated in Latin America (Stotz et al. 1996; Mittermeier et al. 2011; Hedblom and Murgui 2017). Such complexity makes Latin America a focal region for understanding human disturbances and their effects on biodiversity to have a solid foundation to establish evidence-based decision-making.

1.3 Bird Responses to Urbanization

Together with plants, birds are the most studied wildlife group by urban ecologists (McKinney 2008). Such bias occurs for several reasons, of which the following head the list: (i) birds are diverse (Jetz et al. 2012; Navarro-Sigüenza et al. 2014); (ii) form complex communities within urban systems (Blair 1996; Tratalos et al. 2007); (iii) are relatively easy to survey (Ralph et al. 1996); and (iv) respond to modifications in their habitats (Marzluff et al. 2001; Chace and Walsh 2006; Aronson et al. 2014). Since the 1940s, with Pitelka's (1942) report of the birds of a resort village in the California North Coast (United States), urban bird studies have populated the scientific literature, rising importantly after the late 1990s, and reaching a peak in 2012.

With the publication of the iconic book of urban bird studies edited by John M. Marzluff et al. (2001), urban ornithologists have identified several patterns (reviewed by Chace and Walsh 2006 and Evans et al. 2009). For example, bird species richness tends to decrease with urbanization (although it sometimes peaks at intermediate levels of urbanization intensity; Lepczyk et al. 2008), and total bird abundance tends to increase with urbanization, due to the large numbers of a few abundant generalist, often exotic and/or invasive, species. Most recently, Marzluff (2016) reviewed urban bird publications since 2006, recognizing that although some mechanistic processes related to well-studied patterns are being assessed in developed countries (e.g., demographic responses to urbanization), there still exists an important bias toward pattern studies in developing regions, such as Latin America.

Urban bird studies from Latin America have been reviewed in the past, both regionally and nationally (i.e., Colombia; Delgado-V and Correa-H 2013). González-Urrutia (2009) performed a first regional review, focusing on 72 studies, and concluded that urban bird ecology in Latin America was still incipient. In 2011, Ortega-Álvarez and MacGregor-Fors (2011a) performed an in-depth review of 84 studies, starting with Fonaroff's (1974) publication of the urban birds of Port of Spain, Trinidad and Tobago. Recognizing that some general patterns seem to hold in Latin America when contrasted with studies from developed temperate countries, Ortega-Álvarez and MacGregor-Fors (2011a, b) pointed out some differences that had been recorded in the region (e.g., higher species richness recorded in urban Latin America when contrasted with other regions; no evidence of intermediate urbanization scenarios under which avian species richness peak; the important presence of insectivore species in cities across Latin America). Such differences highlight the need to better understand the response of birds to urbanization in a rapidly urbanizing region to add to the global comprehension of this complex phenomenon.

1.4 Avian Studies in Urban Latin America

In this book, we have gathered regional urban bird experts who, refereed by internationally renowned peers, have set the state-of-the-art of urban bird studies in Latin America, contrasted the main retrieved findings with those from other regions across the globe, identified knowledge gaps, and suggested future research needs and directions for each specific topic. This sets the foundation for study of birds in urban Latin America and sheds light on the upcoming steps, hopefully promoting the consolidation of the field in the region. The latter will be possible with the migration toward mechanistic studies that untangle the recorded patterns, and thus generate the required ecological information needed to bridge the gap between evidence-based knowledge and practice in urban systems.

In Chap. 2, Escobar-Ibáñez and MacGregor-Fors set the current status of urban bird studies in Latin America by compiling all available research published since the regional 2011 review (Ortega-Álvarez and MacGregor-Fors 2011a). Focusing their updated review on ecological studies, species lists, and new records, Escobar-Ibáñez and MacGregor-Fors identify several conceptual limitations and spatiotemporal biases that need to be considered with the aim of consolidating the discipline in the region. In Chap. 3, MacGregor-Fors and García-Arroyo focus on bird species richness and composition shifts related to urbanization. Although bird species richness and composition general patterns hold in Latin American cities when compared with the rest of the literature, MacGregor-Fors and García-Arroyo highlight the important number of species that can be recorded within Latin American cities, as well as differences in species composition, mainly assessed using trophic categories. In Chap. 4, Leveau and Zuria review urban avian abundance, demography, and population dynamics, and their urban-related drivers. Based on the response of bird species to urbanization, Leveau and Zuria conclude that, in agreement with studies from other regions, exotic, invasive, and/or generalist species tend to have large populations in urban Latin America, while specialists tend to have low numbers or be absent in urban centers. In Chap. 5, Faggi and Caula focus on the two main infrastructures that make up urban centers: the one comprised by vegetation ('green' infrastructure), and that including man-made structures ('gray' infrastructure). Using this pragmatic framework, Faggi and Caula's synthesis reflects the tight relationship between specific man-made structure and many urban-related bird species, while identifying several 'green' traits that tend to relate with more complex and diverse urban avian communities. In Chap. 6, González-Lagos and Quesada examine the behavioral responses of birds to urbanization. Their analysis focuses on the ethological adjustments to habitat loss and fragmentation, as well as resource availability, pollution (namely noise and artificial night lighting), species interactions, and tolerance to human activities. In Chap. 7, Santiago-Alarcon and Delgado-V summarize the global and regional information regarding urban threats for birds. Covering several major topics (i.e., cat predation, collisions with building structures and vehicles, parasitism), Santiago-Alarcon and Delgado-V identify the crucial need to focus on the effect that urban threats have on Latin American birds due to a concerning lack of studies of most of the evaluated topics. In Chap. 8, Piratelli et al. recognize the general dearth of studies focused on urban bird conservation, with Latin America not being an exception. Although there is evidence in nonpublished and/or confidential technical reports, there is a crucial need to approach the conservation of urban biodiversity through the use of charismatic and informative groups such as birds, where building bird-friendly environments can translate in higher ecological quality, livable, and sustainable cities.

Finally, in Chap. 9, MacGregor-Fors et al. wrap up the book underlining the main findings and conclusions, identifying general knowledge gaps and future research directions. This final chapter pinpoints issues that require urgent attention to enrich future urban bird ecology studies in Latin America, specifically: (i) to consolidate our comprehension of the ecological patterns and related mechanistic processes to understand the response of birds to urbanization, from genes to landscapes; (ii) to standardize the definition of 'urban' and urbanization intensity (including urban–nonurban gradients), as well as urban land uses; (iii) to address the role of the diverse array of spatial (e.g., city size, location) and site-specific variables (e.g., vegetation cover, plant diversity, urbanization intensity), which characterize Latin American cities, on the response of birds to urbanization; (iv) to reevaluate the experimental design and analytical frameworks used for urban bird studies, such as the use of citywide surveys and the inclusion of the physical and social spheres in the interpretation of the studied ecological patterns and processes; and (v) to reexamine several conceptual and analytical approaches in the assessment of biotic homogenization related to urbanization.

Acknowledgments The authors are deeply thankful to Mark Goddard and Rafael Rueda Hernández for their valuable comments and suggestions that improved the clarity of this chapter, and Michelle García-Arroyo for her support in drawing Fig. 1.1. JFE-I acknowledges the scholarship and financial support provided by the National Council of Science and Technology (CONACYT 366146), as well as the Graduate School of the Instituto de Ecología, A.C. (INECOL).

References

Alberti M, Marzluff JM, Shulenberger E et al (2003) Integrating humans into ecology: opportunities and challenges for studying urban ecosystems. Bioscience 53:1169–1179

Aronson MFJ, La Sorte FA, Nilon CH et al (2014) A global analysis of the impacts of urbanization on bird and plant diversity reveals key anthropogenic drivers. Proc R Soc B Biol Sci 281:20133330

Barbosa O, Tratalos J, Armsworth P et al (2007) Who benefits from access to green space? A case study from Sheffield, UK. Landsc Urban Plan 83:187–195

Bazzaz FA (1990) The response of natural ecosystems to the rising global CO2 levels. Annu Rev Ecol Syst 21:167–196

Berkowitz AR, Nilon CH, Hollweg KS (eds) (2003) A new frontier for science and education. Springer, New York

Blair RB (1996) Land use and avian species diversity along an urban gradient. Ecol Appl 6:506–519

Chace JF, Walsh JJ (2006) Urban effects on native avifauna: a review. Landsc Urban Plan 74:46–69

Cincotta RP, Wisnewski J, Engelman R (2000) Human population in the biodiversity hotspots. Nature 404:990–992

Crutzen PJ, Stoermer EF (2000) The Anthropocene. IGBP Newsl 41:17–18

Czech B, Krausman PR, Devers PK (2000) Economic associations among causes of species endangerment in the United States. Bioscience 50:593–601

Delgado-Velez CA, Correa-Hernandez JC (2013) Estudios ornitológicos urbanos en Colombia: Revisión de literatura. Ing Cienc 9:215–236

Eldredge N, Horenstein N (2014) Concrete jungle: New York City and our last best hope for a sustainable future. University of California Press, Oakland

Evans KL, Newson SE, Gaston KJ (2009) Habitat influences on urban avian assemblages. Ibis 151:19–39

Fonaroff LS (1974) Urbanization, birds and ecological change in northwestern Trinidad. Biol Conserv 6:258–262

González-Urrutia M (2009) Avifauna urbana en América Latina: Estudio de casos. Gestión Ambiental 17:55–68

Grimm NB, Faeth SH, Golubiewski NE et al (2008) Global change and the ecology of cities. Science 319:756–760

Hedblom M, Murgui E (2017) Urban bird research in a global perspective. In: Murgui E, Hedblom M (eds) Ecology and conservation of birds in urban environments. Springer, Cham, pp 3–10

Jetz W, Thomas GH, Joy JB et al (2012) The global diversity of birds in space and time. Nature 491:444–448

Lepczyk CA, Flather CH, Radeloff VC et al (2008) Human impacts on regional avian diversity and abundance. Conserv Biol 22:405–416

Liu Z, He C, Zhou Y et al (2014) How much of the world's land has been urbanized, really? A hierarchical framework for avoiding confusion. Landsc Ecol 29:763–771

Marzluff JM (2016) A decadal review of urban ornithology and a prospectus for the future. Ibis 159:1–13

Marzluff JM, Bowman R, Donnelly R (eds) (2001) Avian ecology and conservation in an urbanizing world. Springer, Norwell

Maxwell SL, Fuller RA, Brooks TM (2016) The ravages of guns, nets and bulldozers. Nature 536:143–145

McDonnell MJ (2011) The history of urban ecology: an ecologist's perspective. In: Niemelä J, Breuste JH, Elmqvist T et al (eds) Urban ecology: patterns, processes, and applications. Oxford University Press, Oxford, pp 5–13

McDonnell MJ, MacGregor-Fors I (2016) The ecological future of cities. Science 352:936–938

McKinney ML (2006) Urbanization as a major cause of biotic homogenization. Biol Conserv 127:247–260

McKinney ML (2008) Effects of urbanization on species richness: a review of plants and animals. Urban Ecosyst 11:161–176

Mittermeier RA, Robles-Gil P, Mittermeier CG (eds) (1997) Megadiversity: earth's biologically wealthiest nations. CEMEX/Agrupación Sierra Madre, Mexico City

Mittermeier RA, Turner WR, Larsen FW et al (2011) Global biodiversity conservation: the critical role of hotspots. In: Zachos FE, Habel JC (eds) Biodiversity hotspots. Springer, Heidelberg, pp 3–22

Moreno CE, Sánchez-Rojas G, Pineda E et al (2007) Shortcuts for biodiversity evaluation: a review of terminology and recommendations for the use of target groups, bioindicators and surrogates. Int J Environ Health 1:71–86
Navarro-Sigüenza AG, Rebón-Gallardo MF, Gordillo-Martínez A et al (2014) Biodiversidad de aves en México. Rev Mex Biodivers 85:476–495
Niemelä J (1999) Ecology and urban planning. Biodivers Conserv 8:119–131
Nilon CH, Warren PS, Wolf J (2011) Baltimore birdscape study: identifying habitat and land-cover variables for an urban bird-monitoring project. Urban Habitats 6
Olson DM, Dinerstein E (2002) The Global 200: priority ecoregions for global conservation. Ann Mo Bot Gard 89:199–224
ONU-Habitat (2012) Estado de las ciudades de América Latina y el Caribe. Rumbo a una nueva transición urbana. Naciones Unidas, Rio de Janeiro
Ortega-Álvarez R, MacGregor-Fors I (2011a) Dusting-off the file: a review of knowledge on urban ornithology in Latin America. Landsc Urban Plan 101:1–10
Ortega-Álvarez R, MacGregor-Fors I (2011b) Spreading the word: the ecology of urban birds outside the United States, Canada, and Western Europe. Auk 128:415–418
Pauchard A, Barbosa O (2013) Regional assessment of Latin America: rapid urban development and social economic inequity threaten biodiversity hotspots. In: Elmqvist T, Fragkias M, Goodness J et al (eds) Urbanization, biodiversity and ecosystem services: challenges and opportunities. Springer, Dordrecht, pp 598–608
Pickett STA, Burch WR Jr, Grove J (1999) Interdisciplinary research: maintaining the constructive impulse in a culture of criticism. Ecosystems 2:302–307
Pickett STA, Cadenasso ML, Grove JM et al (2011) Urban ecological systems: scientific foundations and a decade of progress. J Environ Manag 92:331–362
Pickett STA, Cadenasso ML, Childers DL et al (2016) Evolution and future of urban ecological science: ecology in, of, and for the city. Ecosyst Health Sustain 2:e01229
Pitelka FA (1942) High population of breeding birds within an artificial habitat. Condor 44:172–174
Ralph CJ, Geupel GR, Pyle P et al (1996) Manual de métodos de campo para el monitoreo de aves terrestres. Pacific Southwest Research Station, Forest Service, U.S. Department of Agriculture, California
Reyes S, Figueroa I (2010) Distribución, superficie y accesibilidad de las áreas verdes en Santiago de Chile. EURE 26:89–110
Santandreu A, Gómez Perazzoli A, Dubbeling M (2002) Biodiversity, poverty and urban agriculture, in Latin America. Urban Agric Mag 6:9–11
Seto KC, Fragkias M, Güneralp B et al (2011) A meta-analysis of global urban land expansion. PLoS One 6:e23777
Steffen WL, Sanderson A, Tyson PD et al (eds) (2005) Global change and the earth system: a planet under pressure. Springer, New York
Steffen W, Grinevald J, Crutzen P et al (2011) The Antropocene: conceptual and historical perspectives. Phil Trans R Soc A 369:842–867
Steffen W, Broadgate W, Deutsch L et al (2015) The trajectory of the Anthropocene: the great acceleration. Anthropocene Rev 2:81–98
Stotz DF, Fitzpatrick JW, Parker TA et al (eds) (1996) Neotropical birds: ecology and conservation. University of Chicago Press, Chicago
Tratalos J, Fuller RA, Evans KL et al (2007) Bird densities are associated with household densities. Glob Chang Biol 13:1685–1695
Tzoulas K, Korpela K, Venn S et al (2007) Promoting ecosystem and human health in urban areas using green infrastructure: a literature review. Landsc Urban Plan 81:167–178
United Nations (2014) World urbanization prospects: the 2014 revision, highlights. Department of Economic and Social Affairs, Population Division ST/ESA/SER.A/352
United Nations (2015a) World urbanization prospects: the 2014 revision. Department of Economic and Social Affairs, Population Division ST/ESA/SER.A/366

United Nations (2015b) World population 2015. Population Division Department of Economic and Social Affairs United Nations, New York

Venter O, Sanderson EW, Magrach A et al (2016) Sixteen years of change in the global terrestrial human footprint and implications for biodiversity conservation. Nat Commun 7:12558

Vitousek PM, Mooney HA, Lubchenco J et al (1997) Human domination of Earth's ecosystems. Science 277:494–499

Waters C, Zalasiewicz J, Summerhayes C et al (2016) The Anthropocene is functionally and stratigraphically distinct from the Holocene. Science 351:aad2622

Wigginton NS, Fahrenkamp-Uppenbrink J, Wible B et al (2016) Cities are the future. Science 352:904–905

Chapter 2
What's New? An Updated Review of Avian Ecology in Urban Latin America

Juan F. Escobar-Ibáñez and Ian MacGregor-Fors

Abstract Among human activities, urbanization represents one of the most worrisome for biodiversity due to the intensity and long-term effects in the place where a city develops, as well as their indirect effect in its area of influence. Birds are one of the most studied groups to assess the impacts that urbanization has on biodiversity, but there is a lag in the knowledge of highly biodiverse regions like Latin America. This region is the most diverse in avian species, but at the same time is one of the most affected by poverty, social inequality, and population growth, representing a priority for studying urban bird responses to urbanization. In this chapter, we compare results of studies of urban birds in Latin American cities published in recent years with an earlier regional review. We found more than one-third of publications in the last six years than in the previous 35 years; however, there is still a bias in the knowledge toward most populated countries, while urban Central America remains understudied. Although there is a lack of information on the mechanistic processes molding urban avian communities, there are important advances in ecological topics, as novel resource use, urban noise, and urban avian diseases. We urge scientists of Latin American latitudes to join efforts in the understanding of avian communities in urban areas. Results of evidence-based studies ought to be communicated to decision-makers to generate strategies that improve urban management and planning.

Keywords Bird diversity • Literature • Tropics • Urban ecology • Urban gradients

2.1 Urbanization, Birds, and Latin America

As highlighted in Chap. 1, human activities are leading ecosystem changes throughout the world, causing unprecedented impacts on biodiversity and human well-being (Mittermeier et al. 2011; Venter et al. 2016). Among human-related ecological

J.F. Escobar-Ibáñez (✉) • I. MacGregor-Fors
Red de Ambiente y Sustentabilidad, Instituto de Ecología, A.C.
Carretera antigua a Coatepec 351, El Haya, 91070 Xalapa, Veracruz, Mexico
e-mail: juanfer.escobarib@gmail.com

I. MacGregor-Fors, J.F. Escobar-Ibáñez (eds.), *Avian Ecology in Latin American Cityscapes*, DOI 10.1007/978-3-319-63475-3_2

disturbances, urbanization has been identified as one of the most worrisome due to the intensive and long-term modifications that occur where it takes place, as well as the effects beyond its limits (Kennedy et al. 2011). In fact, urbanization has been related to several components of global change (e.g., climate change, biological invasions; Grimm et al. 2008). Thus, it is not surprising that it has been identified as a major driver of species endangerment (Czech et al. 2000; Maxwell et al. 2016).

For the past decades, ecologists have assessed anthropogenic impacts through measurements of biodiversity (Magurran and McGill 2011). Given the high variability and nature among wildlife groups, it has been suggested to follow a 'biodiversity indicator' approach (also known as 'bioindicator') to assess such ecological effects (Moreno et al. 2007). Among 'bioindicators', birds have been widely used because they have high taxonomic diversity (Jetz et al. 2012), represent one of the best known wildlife groups (Schulze et al. 2004), are relatively easily to survey (Ralph et al. 1996), and quickly respond to modifications in their habitats (Marzluff et al. 2001; Aronson et al. 2014). It is not surprising that, based on all of the above factors, birds have been the most studied group in urban centers (McKinney 2006, 2008; Marzluff 2016). Results of studies focused on urban birds, which are heavily biased toward temperate northern developed regions (e.g., United States, Western Europe; Marzluff 2001; Evans et al. 2009; Ortega-Álvarez and MacGregor-Fors 2011a; Hedblom and Murgui 2017), have established some generalizable and consistent patterns. Yet, differences have been reported in understudied regions, including the tropics and subtropics (Ortega-Álvarez and MacGregor-Fors 2011a).

Among tropical and subtropical regions, Latin America has recently gained presence in the urban bird literature (Ortega-Álvarez and MacGregor-Fors 2011a, b). It is noteworthy to underline that this region shelters an important proportion of global avifauna (BirdLife International 2013), encompassing important hotspots for birds (Mittermeier et al. 2011), which is currently exposed to high urbanization rates (United Nations 2015). This makes Latin America a particularly interesting region to further study avian patterns and processes related to urbanization. Insights from this region would not only strengthen our level of comprehension of the response of birds to this complex human disturbance, but will also provide evidence-based information that could create a solid foundation of urban ecology as well as to endorse and promote novel public policies and strategies in our journey of creating biodiverse, sustainable, and livable cities (McDonnell and MacGregor-Fors 2016).

Taking Fonaroff's (1974) pioneer study of avian ecology in northwestern Trinidad as a starting point of bird knowledge in urban Latin America, studies have risen importantly in the last decades. In this chapter, we consider two previous reviews focused on the development of avian studies in urban Latin America as groundwork (González-Urrutia 2009; Ortega-Álvarez and MacGregor-Fors 2011b), and performed an update of the related literature to set a general state-of-the-art on the topic. Finally, we pinpoint some general research gaps and future directions for the discipline, which are refined by topic in each thematic chapter of this book (Chaps. 3–8) and synthesized in the final one (Chap. 9).

2.2 What's New? An Updated Review on Bird Studies in Urban Latin America

To gather as many studies as possible that were published after those considered by previous reviews (González-Urrutia 2009; Ortega-Álvarez and MacGregor-Fors 2011b), we performed an intensive search in Google Scholar (www.scholar.google.com) and all Web of Science (www.webofknowledge.com) databases included in our institutional subscription (i.e., Web of Science Core Collection, Derwent Innovations Index, KCI-Korean Journal Database, Russian Science Citation Index, SciELO Citation Index) seeking for publications from 2009 to 2015, the time period since the last review. In our searches, we used keyword combinations including 'urban' AND 'bird,' followed by the names of 47 Latin American countries (i.e., Anguilla, Antigua and Barbuda, Argentina, Aruba, Bahamas, Barbados, Belize, Bermuda, Bolivia, Brazil, British Virgin Islands, Cayman Islands, Chile, Colombia, Costa Rica, Cuba, Dominica, Dominican Republic, Ecuador, El Salvador, Grenada, Guadeloupe, Guatemala, Guiana, French Guiana, Haiti, Honduras, Jamaica, Martinique, Mexico, Montserrat, Netherlands Antilles, Nicaragua, Panama, Paraguay, Peru, Puerto Rico, Saint Barthélemy, Saint Kitts and Nevis, Saint Lucia, Saint Martin, Saint Vincent and the Grenadines, Suriname, Trinidad and Tobago, Turks and Caicos Islands, Uruguay, and Venezuela).

After carefully reviewing the retrieved set of publications from all used search engines and selecting those that focused on bird studies in urban Latin America, 128 remained (see Appendix 2.1 for references of the publications that were not cited within the chapter). Taking the exhaustive review of 2011 as baseline, which gathered a total of 84 publications in a period of 35 years, the retrieved set of studies represents ~ 9 times higher average annual publication rate. Regarding the geographical distribution of the assessed publications, four countries accounted for ~87% of the articles. Although the same four countries were the best represented in the 2011 review and our update, with Brazil heading the list, Mexico outnumbered the studies from Argentina, and Colombia remained in the same position (Fig. 2.1). It is noteworthy that some countries in which studies were found for previous reviews were absent in our search (French Guiana, Trinidad and Tobago, Panama), while countries as Barbados, Guatemala, Bolivia, and Uruguay recently joined, adding knowledge to our comprehension of avian response to urbanization.

To have a comparable dataset of publications, we used a slightly modified set of categories to those used by Ortega-Álvarez and MacGregor-Fors (2011b) to group the reviewed publications: ecology, species lists, and new records. We did not include a conservation category due to the general lack of conservation-oriented studies in our search (~2%). For further information on the conservation topic, Piratelli et al. (Chap. 8) consider the related literature focused on the reported conservation implications and actions in urban Latin America, as well as the future directions in generating bird-friendly cities. After assigning categories to the set of assessed publications, it was evident that those focused on ecological patterns were still dominant in the literature, represented by ~79% of the retrieved studies, followed by species lists/new records (~19%; Fig. 2.2).

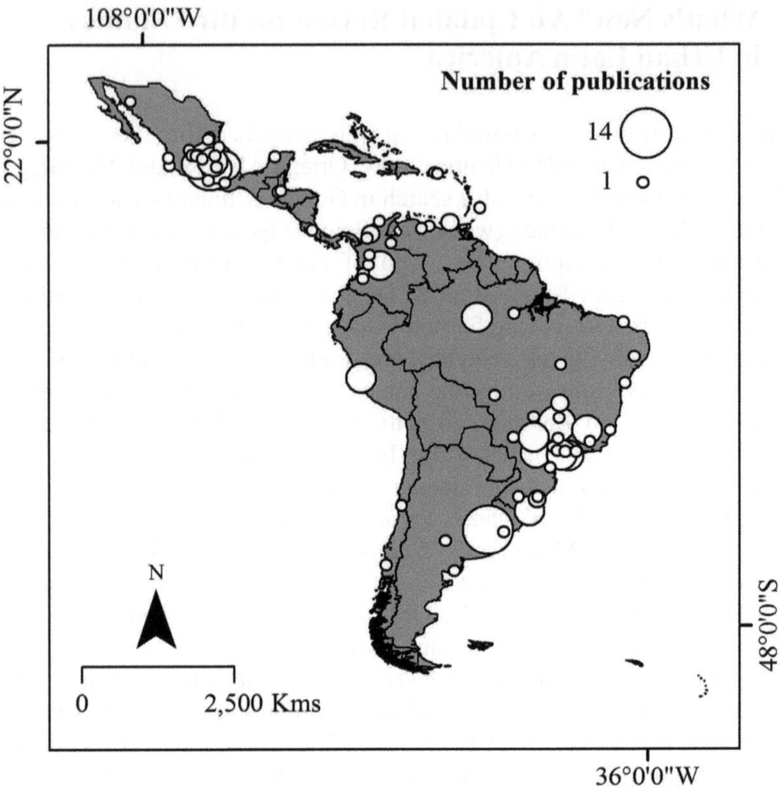

Fig. 2.1 Number of urban bird studies in Latin America by city

2.2.1 Ecological Patterns and Processes

Ecological studies have long represented the majority of urban bird studies in Latin America (Ortega-Álvarez and MacGregor-Fors 2011b); yet, in our updated review, we found a proportional increase in relation to the total retrieved studies, passing from 63% to 78%. Along with such an increase, we identified some specific topics that, as far as we know, had not been previously assessed in the region, including urban bird bioacoustics and parasitism (see Chap. 7). To analyze the retrieved publications, we grouped them in eight main topics that describe their framework and/or aims, as follows: urban gradients, comparisons among different environmental conditions, local traits, regional traits, novel resource use, bird–plant interactions, behavior, noise, and urban avian diseases.

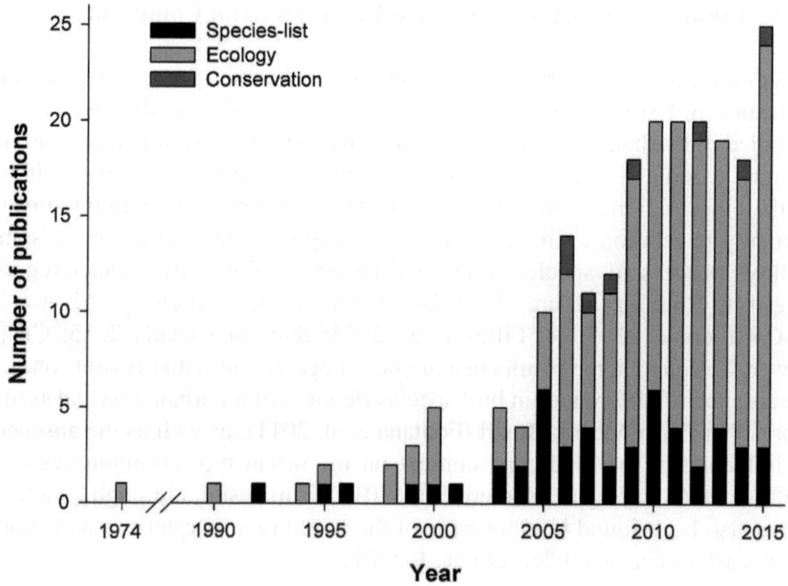

Fig. 2.2 Number of urban bird studies in Latin America by topic and year

2.2.1.1 Urban Gradients

The use of urban gradients is still widely used in the region (Chaps. 3 and 4; also see Chap. 9 for a detailed discussion on the methodological assessment of urban gradients). Some of the general ecological patterns that are consistent with those identified across the globe (Chace and Walsh 2006; Evans et al. 2009) are: (i) decrease in species richness and increase in abundance with urbanization (MacGregor-Fors and Schondube 2012; Toledo et al. 2012; MacGregor-Fors et al. 2015; Silva et al. 2015); (ii) increase in the abundance of a few species that mold the abundance distribution of urban bird communities (Fontana et al. 2011; Toledo et al. 2012); and (iii) increase of compositional similarity in highly urbanized areas (Filloy et al. 2015; Sanz and Caula 2015). One interesting pattern that falls outside of those previously reported was found in La Paz, Bolivia, where bird abundance decreased in highly urbanized areas (Villegas and Garitano-Zavala 2010). Although this pattern was previously reported in City of Olongapo (Philippines; Posa and Sodhi 2006), the study from Bolivia adds evidence for the occurrence of this pattern in tropical areas. Regarding urban to nonurban comparisons, which should not be considered a gradient, but a comparison between different conditions (see Sect. 2.2.1.2, this chapter, as well as Chap. 9 for a detailed discussion), new noticeable results are: the presence of higher concentrations of heavy metals in the feathers of birds from urban areas (Nava-Diaz et al. 2015) and dietary shifts of the Barn Owl (*Tyto alba*) in urban areas when compared to nonurban ones (Teta et al. 2012).

2.2.1.2 Comparisons Among Different Environmental Conditions

Another commonly used approach to assess the response of avian communities to urbanization in Latin America is contrasting urban and surrounding areas. Despite the studied nonurban environments being diverse (e.g., shrublands, grasslands, croplands, woodlands, riparian systems), results are highly consistent with those from the other regions of the globe. For instance, urban avian communities differ importantly in relation to those from surrounding systems, suggesting a 'semipermeability' process of species filtering (MacGregor-Fors 2010; MacGregor-Fors et al. 2011b; Ramírez-Albores 2013; Domínguez-López and Ortega-Álvarez 2014; Puga-Caballero et al. 2014; Filloy et al. 2015; Sanz and Caula 2015; Chap. 6). However, we found some results that are not in agreement with previous ones, such as the absence of differences in bird species richness of an urban area and surrounding forests in Porto Alegre, Brazil (Fontana et al. 2011), as well as the absence of a clear influence of the landscape context on the urban bird communities of periurban Mexico City (Puga-Caballero et al. 2014). This result, although counterintuitive, has also been found in other parts of the world (e.g., western France, northern Finland, eastern Canada; Clergeau et al. 2001).

2.2.1.3 Local Environmental Predictors

At the local scale, most studies report effects of some key environmental characteristics on bird species richness, including vegetation diversity, structure, and composition, as well as proxies of urbanization intensity (see Chaps. 3, 5, and 9 for further methodological discussions). Results of many studies carried out in urban Latin America are in agreement with those from across the globe, finding positive relationships between vegetation complexity (e.g., Toledo et al. 2012; Leveau 2013) and vegetation cover (e.g., Villegas and Garitano-Zavala 2010; Nolazco 2012), and negative relationships with percentage of impervious surface (e.g., Villegas and Garitano-Zavala 2010; Reis et al. 2012; Sanz and Caula 2015), noise (e.g., Fontana et al. 2011), and bird species richness (see Chap. 3 for further details on this topic). However, avian groups have been shown to respond differently to urbanization depending on their specific requirements (Blair 1996; Rebolo and Fiorini 2010; MacGregor-Fors and Schondube 2011; Toledo et al. 2012; see Chap. 6 for a behavioral perspective). Surprisingly, one study found negative relationships between native bird species richness and vegetation richness (Silva et al. 2015), although the authors warn about the low explanatory power of the model (16%).

2.2.1.4 Regional Scale Predictors

Regional scale drivers of species richness, such as species–area relationships, are again similar to those identified in global studies. Positive relationships have been recorded in Mexican urban areas between size of human settlements and bird species richness (MacGregor-Fors et al. 2011a), as well as with the size of urban greenspaces (Carbó-Ramírez and Zuria 2011; Charre et al. 2013). Interestingly, a study comparing several urbanized settlements of west-central Mexico found that, on a regional scale, settlement size can override the contributions of built cover, human population density, tree cover, and plant diversity (MacGregor-Fors et al. 2011a). On the other hand, considering another island biogeography concept, 'distance from the source' or 'distance to the city border' has been reported to be negatively related to bird species richness (MacGregor-Fors and Ortega-Álvarez 2011; Silva et al. 2015), but this relationship occurs differently for particular bird groups (e.g., native, exotic, migrants; MacGregor-Fors and Ortega-Álvarez 2011; Charre et al. 2013).

2.2.1.5 Novel Resource Use

The ability of organisms to use new resources in urban areas can determine their prevalence in such novel environments (Chap. 6). We found records of the use of resources like thermal power plants by Black (*Coragyps atratus*) and Turkey Vultures in Brazil (*Cathartes aura*; Freire et al. 2015), as well as open garbage containers by some bird species (e.g., Black Vulture; Novaes and Cintra 2013). Similarly, there are interactions not previously recorded in the literature, either in natural and/or modified environments, such as the consumption of bats and rats by Harris's Hawks (*Parabuteo unicinctus*; Ortega-Álvarez and Calderón-Parra 2014) and Channel-billed Toucans (*Ramphastos vitelinus*; da Silva and Azevedo 2013), respectively. Novel nesting sites have also been recorded in urban Latin America, such as Azure-crowned Hummingbirds (*Amazilia cyanocephala*) nesting in open sky wires (Escobar-Ibáñez and MacGregor-Fors 2016) and Kelp Gulls (*Larus dominicanus*) nesting in fuel tanks (Villavicencio 2014). Also, regarding urban nesting, one study from Mexico City found the use of cigarette butts in the nests of House Sparrows (*Passer domesticus*) and House Finches (*Haemorhous mexicanus*) that reduce parasite loads in the nests (Suárez-Rodríguez et al. 2012), and thus could act as repellent (although recent research suggests that incorporating cigarette butts could increase breeding costs; Suárez-Rodríguez et al. 2017). Another interesting study shows that the threatened Burrowing Parrot (*Cyanoliseus patagonus*) reproduces successfully at highly disturbed sites, such as locations in which waterholes are established and construction materials are extracted in Argentina (Tella et al. 2014).

2.2.1.6 Bird-Plant Interactions

As discussed above, urban vegetation has been shown to be tightly related with urban avian diversity. Bird-plant interaction studies have been increasingly present in the urban ecology literature from Latin America, focused both in 'green' and 'gray' areas (Chap. 5), with an important bias toward nectar and fruit consumption (Gómez-Aíza and Zuria 2010; Leveau and Leveau 2011; Rodrigues and Araujo

2011; Maruyama et al. 2012; Andrade et al. 2013; Ferreira and Consolaro 2013; Santos and Ragusa-Netto 2014). One pioneer study evaluated the relationship between the spectral properties of flowers and their visitors in northeastern São Paulo (Toledo and Donatelli 2010), finding more visitor species in those flowers with long wavelengths, although further studies are needed to fully understand the complex network of interactions in which nectar-flowering plants, birds, and other wildlife groups are implied. Although not necessarily novel, several studies have highlighted the importance of exotic trees providing alternative food resources year-round (Oliveira et al. 2013; Previatto et al. 2013). Remarkably, even large-sized frugivores have been recorded to take advantage of such resources in urban Latin America (e.g., Toco Toucan—*Ramphastos toco*, and Blue-and-yellow Macaw—*Ara ararauna*; Santos and Ragusa-Netto 2013, 2014). Studies have suggested considering this in urban ecological management plans, since large frugivores are one of the most susceptible avian groups to human disturbances (Franz et al. 2010; Lessi et al. 2016). Yet, there is evidence that such a positive effect does not always occur, with exotic plant density also being negatively related with urban avian species richness (Reis et al. 2012).

2.2.1.7 Behavior

Behavioral plasticity is one of the most important factors determining the arrival and establishment of birds in urban areas (see Chap. 6 for a thorough revision on this topic). Recent studies have added to our knowledge on avian responses to urbanization from an ethological perspective, such as urban-related resource use by flycatchers (e.g., Great Kiskadee—*Pitangus sulphuratus*, Tropical Kingbird—*Tyrannus melancholicus*; Martins-Oliveira et al. 2012), the flight initiation distance of Burrowing Owl (*Athene cunicularia*; Rebolo-Ifrán et al. 2015), and bolder and better problem-solving urban wild-caught Barbados Bullfinches (*Loxigilla barbadensis*) when compared to individuals captured in nonurban areas (Audet et al. 2016). Additional intriguing avian behavioral responses have been recently reported in urban Latin American sites, including the commensal association of egrets, kingfishers, and herons foraging with otters in an urban greenspace of southeastern Brazil (D'Angelo and Sazima 2014).

2.2.1.8 Noise

Environmental noise seems to play a major role on the establishment of bird species within urban systems (González-Oreja et al. 2012; Chap. 7). This topic has gained important presence in recent years, with studies identifying urban-related bird species (e.g., House Finch, Vermilion Flycatcher—*Pyrocephalus rubinus*) adjusting their vocalizations to avoid noise interference (Bermúdez-Cuamatzin et al. 2009). We found a binational study focused on Mexican and Brazilian birds testing the vocal flexibility of oscine (songbirds that learn to sing) and suboscine birds (those that lack song learning) in urban systems. Results of the study show that song transmission capacity and ambient acoustics are related for oscines but not for suboscines, suggesting that song learning can be an important mechanism for adaptation in urban areas (Ríos-Chelén et al. 2012).

2.2.1.9 Urban Avian Diseases

Viruses, parasites, bacteria, and fungus are vectors of diverse diseases of ecological and public health concerns (Oliveira et al. 2012; Kading et al. 2013; see Chap. 7 for a thorough review on the topic). There has been an important increase in the number of studies focused on this topic, going from one study identified in the 2011 review (i.e., Monnerat-Nogueira et al. 2005) to 17 studies published in the time frame considered in this review. Interestingly, ~70% of the gathered studies were conducted in Brazil. The vast majority of publications focused on exotic bird diseases, mainly Rock Pigeons—*Columba livia* (de Sousa et al. 2010; de Lima et al. 2011; Caballero et al. 2015; Pérez-García et al. 2015) and House Sparrows (Gondim et al. 2010; Komar et al. 2012), finding *Escherichia coli* in 86% of Rock Pigeon samples (Silva et al. 2009), as well as viruses (e.g., West Nile virus) and fungi (*Cryptococcus* and *Candida* spp.) in 95% of Rock Pigeon samples (Costa et al. 2010). A more ecological study performed in southeastern Mexico shows that urbanization can play a crucial role in the frequency and severity of avian malaria infections for the Chestnut-capped Brushfinch (*Arremon brunneinucha*; Hernández-Lara et al. 2017). Other studies recorded new bacteria in bird species (e.g., *Neospora caninum* in house sparrows, *Dispharynx nasuta* in Picui Ground-Doves, Poxvirus in Barn Owls; Coimbra et al. 2009; Vargas et al. 2011), and even new pathogen genotypes in Eared Doves—*Zenaida auriculata* (i.e., *Toxoplasma gondii* #182; Barros et al. 2014). Of the few publications addressing ecological questions, the main conclusions are: (i) parasite richness and abundance decrease with urbanization (Calegaro-Marques and Amato 2014); (ii) avian host abundance of the West Nile virus increases with urbanization (Morales-Betoulle et al. 2013); and (iii) higher prevalence of parasites in native species when compared to exotics (Lima et al. 2010).

2.2.2 *Species Lists and New Records*

Urban-related species lists and new records continue to be the second most common topic in the urban bird literature from Latin America. Yet, compared with the 2011 review, there has been a reduction in the proportion of publications (from ~32% in 2011 to ~18.5% to 2015). A decrease in this kind of publications and the increase in ecological studies are parts of a transition toward a more comprehensive approach in the region (Table 2.1). As pointed out in the 2011 review, one of the main differences between species lists of temperate urban areas and those from Latin America is the high level of reported avian diversity (Table 2.2). There is an example of a

Table 2.1 New records for birds in urban Latin America between 2009 and 2015

Species name	Common name	City (Country)	Reference
Dendrocygna bicolor	Fulvous Whistling-Duck	Cúcuta (Colombia)	Armesto et al. (2013)
Streptopelia decaocto	Eurasian Collared-Dove	Querétaro (Mexico)	Pineda-López and Malagamba (2011)
Columbina inca	Inca Dove	Ciudad Victoria (Mexico), Jaumave (Mexico)	Rodríguez-Ruiz et al. (2015)
Coccyzus americanus	Yellow-billed Cuckoo	Cúcuta (Colombia)	Armesto et al. (2013)
Pelecanus occidentalis	Brown Pelican	Santarém (Brazil)	Almeida-Santos et al. (2016)
Myiopsitta monachus	Monk Parakeet	Mexico (country), La Plata (Argentina), Pampas grassland ecoregion (Argentina), Buenos Aires (Argentina)	MacGregor-Fors et al. (2011a), Pineda-López and Malagamba (2011), Volpe and Aramburú (2011), Pineda-López et al. (2013), Bucher and Aramburú (2014), Romero et al. (2015)
Forpus conspicillatus	Spectacled Parrotlet	Cúcuta (Colombia)	Armesto et al. (2013)
Vireo philadelphicus	Philadelphia Vireo	Mexico City (Mexico)	Aguilar-Gómez et al. (2015)
Turdus rufopalliatus	Rufous-backed Robin	León (Mexico); Querétaro (Mexico), Central Mexico	Pineda-López and Malagamba (2009), Martínez-Morales et al. (2010), Pineda-López et al. (2013)
Sturnus vulgaris	European Starling	Buenos Aires (Argentina), León (Mexico), Querétaro (Mexico)	Rebolo and Fiorini (2010), Pineda-López and Malagamba (2011), Pineda-López et al. (2013)
Protonotaria citrea	Prothonotary Warbler	Cúcuta (Colombia)	Armesto et al. (2013)
Paroaria nigrogenis	Masked Cardinal	Cúcuta (Colombia)	Armesto et al. (2013)
Zonotrichia leucophrys	White-crowned Sparrow	Mexico City (Mexico)	Aguilar-Gómez et al. (2015)
Quiscalus mexicanus	Great-tailed Grackle	Ciudad Victoria (Mexico), Jaumave (Mexico)	Rodríguez-Ruiz et al. (2015)

Table 2.2 Bird species lists from urban Latin America between 2009 and 2015

City/country	Species richness	Reference
Caracas (Venezuela)	40	Sainz-Borgo (2015)
Puebla (Mexico)	46	Almazán-Núñez and Hinterholzer-Rodríguez (2010)
Universidad Nacional Agraria la Molina, lima (Peru)	48	Castillo-Palacios et al. (2014)
Sorocaba (Brazil)	65	Cruz and Piratelli (2011)
Universidad Autónoma de Guerrero campus Sur, Chilpancingo (Mexico)	76	Castro-Torreblanca and Blancas-Calva (2009)
Pelotas (Brazil)	84	Sacco et al. (2013)
Universidad del Cauca, Popayán (Colombia)	114	Ramírez-Cháves et al. (2010)
Pelotas (Brazil)	114	Franz et al. (2010)
Belo-Horizonte (Brazil)	123	Vasconcelos et al. (2013)
Puerto Vallarta (Mexico)	160	Molina et al. (2012)
Universidad de Magdalena, Santa Marta (Colombia)	186	Strewe et al. (2009)
Pereira (Colombia)	207	Londoño-Betancourth (2011)
Xalapa (Mexico)	242	Ruelas Inzunza and Aguilar Rodríguez (2010)
Xalapa (Mexico)	341	González-García et al. (2016)

neotropical ctiy for which more than 340 bird species have been recorded in the last three decades (i.e., Xalapa, Mexico, ~64 km^2; González-García et al. 2016). Aside from the study of Xalapa and that of the campus of the Pontifícia Universidade Católica de Minas Gerais (performed along three decades; Vasconcelos et al. 2013), most species lists are based on short time windows or comparisons through time (Franz et al. 2010; Sanz et al. 2010; Biamonte et al. 2011; Vasconcelos et al. 2013; Castillo-Palacios et al. 2014). One of the most dramatic results regarding temporal comparisons was that reported by Biamonte et al. (2011) in relation to the urban sprawl of San José, Costa Rica, finding: (i) the local extinction of 32 resident and 34 latitudinal migratory bird species; (ii) decrease in the abundance of resident and migratory species; (iii) increase of species related to open areas; and (iv) decrease of forest-dependent species.

Regarding the identity of species reported in the species lists, we found both similarities and remarkable differences when compared to the 2011 review. There is a general agreement that the best represented groups throughout Latin American cities are flycatchers (e.g., Strewe et al. 2009; Almazán-Núñez and Hinterholzer-Rodríguez 2010; Cruz and Piratelli 2011; Toledo et al. 2012), although sometimes these groups are underrepresented when compared to adjacent nonurban systems (Fontana et al. 2011). While the 2011 review identified emberizids (e.g., sparrow, buntings) as one of the best represented families, we found that warblers and tanagers are often the most represented groups after flycatchers, followed by icterids (e.g., orioles, grackles), emberizids, and hummingbirds (e.g., Caula et al. 2010; Carbó-Ramírez and Zuria 2011; Nolazco 2012; Charre et al. 2013). Detailed analyses focused by country show evident geographical variations. For example, warblers are the best represented group in urban Mexico, followed by flycatchers (Almazán-Núñez and Hinterholzer-Rodríguez 2010; Carbó-Ramírez and Zuria 2011; Molina et al. 2012; Charre et al. 2013; González-García et al. 2014), while flycatchers are the best represented group in Brazil, Colombia, Peru, Venezuela, and Bolivia, followed by tanagers (Castro-Torreblanca and Blancas-Calva 2009; Caula et al. 2010; Ramírez-Cháves et al. 2010; Villegas and Garitano-Zavala 2010; Biamonte et al. 2011; Londoño-Betancourth 2011; Reis et al. 2012; Toledo et al. 2012; Sainz-Borgo 2015).

2.3 Conclusions

Bird studies in urban Latin America have noticeably increased in the last years, providing significant knowledge advances in both well-studied and new topics, including those from countries for which no publications had been identified in previous reviews (i.e., Guatemala, Barbados, Uruguay, Bolivia). Yet, there are some countries that have not continued to contribute with knowledge since the 2011 review (e.g., Trinidad and Tobago, Panama), and many other countries that, as far as we know, have not developed urban avian studies. In this concluding section, we briefly provide a general perspective of the main advances and the future research directions for the study of birds in urban Latin America (see the thematic chapters for specific topic-related advances and future directions; Chaps. 3, 4, 5, 6, 7, and 8).

Our updated review shows that urban avian ecology in Latin America is currently in a transition phase that is moving forward to describe ecological patterns, and even starting to dive into some of the mechanistic processes behind them. Yet, one interesting and worrisome pattern found through our review was a decrease in the proportion of publications of species lists and new records in comparison with the 2011 review. Although it is crucial to generate information regarding the patterns and processes to understand the way in which birds are responding, acclimatizing, and adapting to urbanization in Latin America, basic information is fundamental for further studies. Thus, we recognize the importance of intensifying efforts to keep the generation of basic natural history studies, as well as new records, and species lists together with ecological, genetic, conservation, management, and planning-oriented studies.

Regarding general patterns, few followed a comparative urban–nonurban contrast, finding bird richness decreases with urbanization (e.g., MacGregor-Fors et al. 2011b; Sanz and Caula 2015). However, the kind of ecotone can influence the magnitude of these effects (Puga-Caballero et al. 2014). It is imperative that studies adopt a comparative approach to understand the way in which birds respond to urbanization in the widely diverse set of environmental conditions and bird diversity present throughout Latin America.

We found several studies suggesting that their results show both taxonomic and functional homogenization (e.g., Ortega-Álvarez and MacGregor-Fors 2009; Filloy et al. 2015). However, there are some methodological and conceptual aspects that have not been considered, and we highly recommend interpreting them cautiously (see Chap. 9 for a detailed analysis). Regarding local and regional predictors of bird species richness, we found several similarities to what has been previously reported for the region and others across the globe. For example, vegetation components mold bird diversity within cities at several spatial scales. Although many typical vegetation measures have been assessed (e.g., tree, shrub, and herb cover, DBH, and height; origin, phenology), there are other variables for which we have little information, such as greenspace connectivity, the role of private gardens, green roofs (although scarce in the region), and the bird use of dead trees.

Fortunately, studies focused on some processes of the reported urban bird patterns have started to gain presence in the Latin American bird literature. It is clear that these studies have contributed with pioneer findings, often dealing with acclimatory responses to urbanization that broaden our comprehension of the mechanisms related to some of the described patterns. Yet, these types of studies are scarce and biased toward specific topics. For example, recent studies have assessed the novel use of resources, such as the ability of birds to survive in urban conditions (e.g., Tella et al. 2014), song adjustments to urban noise (e.g., Ríos-Chelén et al. 2012), avian disease diversity related to urbanization (Calegaro-Marques and Amato 2014), spectral properties of flowers with avian feeding visiting rate (Toledo and Donatelli 2010), and the use of artificial resources for nesting (e.g., Suárez-Rodríguez et al. 2012). Interestingly, follow-ups of these studies have untangled important details regarding the main findings. For example, a study assessing the use of cigarette butts by urban birds in Mexico City concluded that while smoked cigarette butts act as ectoparasite repellents, they are also genotoxic, which could represent a negative breeding cost (Suárez-Rodríguez et al. 2017).

Based on our literature review, we identified some general knowledge gaps in relation to our understanding of the effect that urbanization has on birds in Latin America. Such gaps are not only conceptual, but also related with temporal and spatial biases. For example, the time-window of studies is often short, limiting our understanding on the long term, although some studies have addressed temporal changes through decadal comparisons to evaluate avian changes caused by urbanization through time (e.g., Biamonte et al. 2011). Also, the common focus on specific cities, often from the same countries, as well as the use of single-city designs and city comparisons within but not among countries, limits our understanding of the specific factors related with urban bird diversity along the diverse array of urban conditions embedded in a highly diverse cultural region.

Despite the myriad of urban management and planning recommendations based on the assessment of how birds, a well recognized bioindicator, respond to urbanization, many Latin American cities continue to sprawl and be managed without including such knowledge (Silva et al. 2015). Due to the complexity of urban ecosystems, there is an urgent need to include evidence-based ecological knowledge in the urban management, planning, and conservation agendas of the region (Miller and Hobbs 2002; Cestari 2015). Willingness from all the involved stakeholders, from householders to governmental institutions, is key in bridging the gap between evidence-based knowledge and its application (MacGregor-Fors 2015). Undoubtedly, the more we understand the history of our cities, as well as the physical, ecological and social patterns, and related processes, we will be closer to setting balance between our modern urban way of life and the drastic ecological effects it triggers.

Acknowledgments The authors thank Paige S. Warren and Javier Laborde for their constructive and comprehensive comments and suggestions, which undoubtedly enhanced the quality and clarity of this chapter. They are also grateful to Michelle García-Arroyo for drawing Fig. 2.1. JFE-I thanks Anne and Ed Stanek for their hospitality during the writing of part of this chapter and acknowledges the scholarship and financial support provided by the National Council of Science and Technology (CONACYT 366146), as well as the Graduate School of the Instituto de Ecología, A.C. (INECOL).

Appendix 2.1 List of Retrieved Publications in the Search that Were Not Cited in the Chapter

Andery D de A, Ferreira Junior FC, de Araújo AV et al (2013) Health assessment of raptors in triage in Belo Horizonte, MG, Brazil. Rev Bras Ciênc Avíc 15:247–256

Azevedo CS, Silva KS, Ferraz JB et al (2012) Does people's knowledge about an endangered bird species differ between rural and urban communities? The case of the Greater Rhea (*Rhea americana*, Rheidae) in Minas Gerais, Brazil. Rev Bras Ornitol 20:8–18

Boogert NJ, Monceau K, Lefebvre L (2010) A field test of behavioural flexibility in Zenaida Doves (*Zenaida aurita*). Behav Process 85:135–141

Carbó-Ramírez P, González-Arrieta RA, Zuria I (2015) Breeding biology of the Rufous-backed Robin (*Turdus rufopalliatus*) in an urban area outside its original distribution range. Wilson J Ornithol 127:515–521

Carrete M, Tella JL (2011) Inter-individual variability in fear of humans and relative brain size of the species are related to contemporary urban invasion in birds. PLoS One 6:e18859

Carrete M, Tella JL (2013) High individual consistency in fear of humans throughout the adult lifespan of rural and urban burrowing owls. Sci Rep 3:3524

Chávez-Zichinelli CA, MacGregor-Fors I, Rohana PT et al (2010) Stress responses of the House Sparrow (*Passer domesticus*) to different urban land uses. Landsc Urban Plan 98:183–189

Fontoura PM, Orsi ML (2013) Ecological partitioning of three Columbidae species in Northern Paraná, Southern Brazil. Biota Neotrop 13:44–49

González-Oreja JA (2011) Birds of different biogeographic origins respond in contrasting ways to urbanization. Biol Conserv 144:234–242

Jebai GT, Arakaki BR, Silva CAP et al (2013) Análise comparativa da densidade de onze passeriformes em duas áreas urbanas em Londrina, norte do Paraná, Brasil. Rev Bras Ornitol 17:183–186

Johnston López K, Macías Duarte A, Castillo Gámez RA (2015) Urban birds in the Sonoran Desert: Estimating population density from point counts. Huitzil 16:37–47

Laiolo P (2011) Homogenisation of birdsong along a natural–urban gradient in Argentina. Ethol Ecol Evol 23:274–287

Leveau LM, Isla FI, Bellocq MI (2015) Urbanization and the temporal homogenization of bird communities: A case study in central Argentina. Urban Ecosyst 18:1461–1476

MacGregor-Fors I, Morales-Pérez L, Quesada J et al (2010) Relationship between the presence of House Sparrows (*Passer domesticus*) and Neotropical bird community structure and diversity. Biol Invasions 12:87–96

Marques-Santos F, Wischhoff U, Rodrigues M (2014) New insights on the rarity of the vulnerable Cinereous Warbling-finch (Aves, Emberizidae) based on density, home range, and habitat selection. Braz J Biol 74:795–802

Parry L, Barlow J, Pereira H (2014) Wildlife harvest and consumption in Amazonia's urbanized wilderness: Wildlife consumption in urbanized Amazonia. Conserv Lett 7:565–574

Rodriguez-Martínez S, Carrete M, Roques S et al (2014) High urban breeding densities do not disrupt genetic monogamy in a bird species. PLoS One 9:e91314

Rosselli L, Stiles FG (2012a) Wetland habitats of the Sabana de Bogotá Andean highland plateau and their birds. Aquat Conserv 22:303–317

Rosselli L, Stiles FG (2012b) Local and landscape environmental factors are important for the conservation of endangered wetland birds in a high Andean Plateau. Waterbirds 35:453–469

Sazima I (2009) Insect cornucopia: Various bird types prey on the season's first giant cicadas in an urban park in southeastern Brazil. Biota Neotrop 9:259–262

Sazima I, D'Angelo GB (2012) Agonistic interactions between two foraging Anhinga females in Southeastern Brazil. Wilson J Ornithol 124:403–405

Solaro C, Sarasola JH (2015) Nest-spacing, not human presence, influences the breeding of Chimango Caracaras (*Milvago chimango*) in a peri-urban reserve. Emu 115:72–75

Zaldúa N, Hernández D, Rodríguez-Tricot L (2013) Evaluación del uso de cajas nido por parte de aves urbanas. Bol Soc Zool Uruguay 22:39–45

References

Aguilar-Gómez MÁ, Calderón-Parra R, Ortega-Álvarez R (2015) Nuevos pájaros en la ciudad de México: Primeros registros de *Vireo philadelphicus* y *Zonotrichia leucophrys* para el Distrito Federal. Huitzil 16:33–36

Almazán-Núñez RC, Hinterholzer-Rodríguez A (2010) Dinámica temporal de la avifauna en un parque urbano de la ciudad de Puebla, México. Huitzil 11:26–32

Almeida-Santos DA, Ferreira GS, Lopes EV (2016) New record of the Brown Pelican *Pelecanus occidentalis* in continental waters of the Brazilian Eastern Amazon. Rev Bras Ornitol 23:351–353

Andrade PC, Mota JVL, Carvalho AAF (2013) Mutual interactions between frugivorous birds and plants in an urban fragment of Atlantic forest, Salvador, BA. Rev Bras Ornitol 19:12

Armesto LO, Torrado Vargas RA, Esteban JB (2013) Registro de cinco especies de aves poco conocidas para norte de Santander, Colombia. Acta Biol Colomb 18:199–204

Aronson MFJ, La Sorte FA, Nilon CH et al (2014) A global analysis of the impacts of urbanization on bird and plant diversity reveals key anthropogenic drivers. Proc R Soc Lond B Biol Sci 281:20133330

Audet J-N, Ducatez S, Lefebvre L (2016) The town bird and the country bird: problem solving and immunocompetence vary with urbanization. Behav Ecol 27:637–644

Barros LD, Taroda A, Zulpo DL et al (2014) Genetic characterization of *Toxoplasma gondii* isolates from Eared Doves (*Zenaida auriculata*) in Brazil. Rev Bras Parasitol Vet. 23:443–448

Bermúdez-Cuamatzin E, Ríos-Chelén AA, Gil D et al (2009) Strategies of song adaptation to urban noise in the house finch: syllable pitch plasticity or differential syllable use? Behaviour 146:1269–1286

Biamonte E, Sandoval L, Chacón E et al (2011) Effect of urbanization on the avifauna in a tropical metropolitan area. Landsc Ecol 26:183–194

BirdLife International (2013) Birds are found almost everywhere in the world, from the poles to the equator. http://datazone.birdlife.org/sowb/casestudy/birds-are-found-almost-everywhere-in-the-world-from-the-poles-to-the-equator. Accessed 20 Feb 2017

Blair RB (1996) Land use and avian species diversity along an urban gradient. Ecol Appl 6:506–519

Bucher EH, Aramburú RM (2014) Land-use changes and Monk Parakeet expansion in the Pampas grasslands of Argentina. J Biogeogr 41:1160–1170

Caballero M, Rivera I, Jara LM et al (2015) Isolation and molecular identification of potentially pathogenic *Escherichia coli* and *Campylobacter jejuni* in feral pigeons from an urban area in the city of Lima, Peru. Rev Inst Med Trop São Paulo 57:393–396

Calegaro-Marques C, Amato SB (2014) Urbanization breaks up host-parasite interactions: a case study on parasite community ecology of Rufous-bellied Thrushes (*Turdus rufiventris*) along a rural-urban gradient. PLoS One 9:e103144

Carbó-Ramírez P, Zuria I (2011) The value of small urban greenspaces for birds in a Mexican city. Landsc Urban Plan 100:213–222

Castillo-Palacios L, Castañeda-Córdova L, Quinteros-Carlos Z (2014) Aves del campus de la Universidad Nacional Agraria La Molina (Lima-Perú) – Una revisión de su abundancia, distribución y diversidad desde 1992 al 2010. Ecol Apl 13:117–128

Castro-Torreblanca M, Blancas-Calva E (2009) Aves de Ciudad Universitaria campus sur de la Universidad Autónoma de Guerrero, Chilpancingo, Guerrero, México. Huitzil 15:82–92

Caula S, Giner S, De Nóbrega JR (2010) Aves urbanas: Un estudio comparativo en dos parques tropicales con diferente grado de intervención humana (Valencia, Venezuela). Faraute 5:1–13

Cestari C (2015) Coexistence between Nearctic-Neotropical migratory shorebirds and humans on urban beaches of the Southern Hemisphere: a current conservation challenge in developing countries. Urban Ecosyst 18:285–291

Chace JF, Walsh JJ (2006) Urban effects on native avifauna: a review. Landsc Urban Plan 74:46–69

Charre GM, Hurtado AZ, Néve G et al (2013) Relationship between habitat traits and bird diversity and composition in selected urban green areas of Mexico City. Ornitol Neotrop 24:279–297

Clergeau P, Jökimaki J, Savard J (2001) Are urban bird communities influenced by the bird diversity of adjacent landscapes? J Appl Ecol 38:1122–1134

Coimbra MAA, Mascarenhas CS, Krüger C et al (2009) Helminths parasitizing *Columbina picui* (Columbiformes: Columbidae) in Brazil. J Parasitol 95:1011–1012

Costa AKF, Sidrim JJC, Cordeiro RA et al (2010) Urban pigeons (*Columba livia*) as a potential source of pathogenic yeasts: a focus on antifungal susceptibility of cryptococcus strains in Northeast Brazil. Mycopathologia 169:207–213

Cruz BB, Piratelli AJ (2011) Avifauna associada a um trecho urbano do Rio Sorocaba, Sudeste do Brasil. Biota Neotropica 11:255–264

Czech B, Krausman PR, Devers PK (2000) Economic associations among causes of species endangerment in the United States. Bioscience 50:593–601

da Silva JN, Azevedo CS (2013) *Rattus rattus* (Mammalia: Rodentia) predation by *Ramphastos vitellinus* (Aves: Ramphastidae) in Santa teresa municipality, Espírito Santo, Brazil. Rev Bras Ornitol 20:2

D'Angelo GB, Sazima I (2014) Commensal association of piscivorous birds with foraging otters in southeastern Brazil, and a comparison with such a relationship of piscivorous birds with cormorants. J Nat Hist 48:241–249

de Lima VY, Langoni H, da Silva AV et al (2011) *Chlamydophila psittaci* and *Toxoplasma gondii* infection in pigeons (*Columba livia*) from São Paulo state, Brazil. Vet Parasitol 175:9–14

de Sousa E, Júnior AB, Pinto AA et al (2010) Prevalence of *Salmonella* spp. antibodies to *Toxoplasma gondii*, and Newcastle disease virus in feral pigeons (*Columba livia*) in the city of Jaboticabal, Brazil. J Zoo Wildl Med 41:603–607

Domínguez-López ME, Ortega-Álvarez R (2014) The importance of riparian habitats for avian communities in a highly human-modified Neotropical landscape. Rev Mex Biodivers 85:1217–1227

Escobar-Ibáñez JF, MacGregor-Fors I (2016) Peeking into the past to plan the future: assessing bird species richness in a neotropical city. Urban Ecosyst 19:657–667

Evans KL, Newson SE, Gaston KJ (2009) Habitat influences on urban avian assemblages. Ibis 151:19–39

Ferreira MC, Consolaro HN (2013) Fenologia e síndromes de polinização e dispersão de espécies de sub-bosque em um remanescente florestal urbano no Brasil central. Biosci J 29:1708–1720

Filloy J, Grosso S, Bellocq MI (2015) Urbanization altered latitudinal patterns of bird diversity-environment relationships in the southern Neotropics. Urban Ecosyst 18:777–791

Fonaroff LS (1974) Urbanization, birds, and ecological change in northwestern Trinidad. Biol Conserv 6:258–262

Fontana CS, Burger MI, Magnusson WE (2011) Bird diversity in a subtropical South-American City: effects of noise levels, arborisation and human population density. Urban Ecosyst 14:341–360

Franz I, Cappelatti L, Barros MP (2010) Bird community in a forest patch isolated by the urban matrix at the Sinos River basin, Rio Grande do Sul state, Brazil, with comments on the possible local defaunation. Braz J Biol 70:1137–1148

Freire DA, Gomes FBR, Cintra R et al (2015) Use of thermal power plants by New World vultures (Cathartidae) as an artifice to gain lift. Wilson J Ornithol 127:119–123

Gómez-Aíza L, Zuria I (2010) Aves visitantes a las flores del maguey (*Agave salmiana*) en una zona urbana del centro de México. Ornitol Neotrop 21:17–30

Gondim LSQ, Abe-Sandes K, Uzêda RS et al (2010) *Toxoplasma gondii* and *Neospora caninum* in sparrows (*Passer domesticus*) in the Northeast of Brazil. Vet Parasitol 168:121–124

González-García F, Straub R, García JAL et al (2014) Birds of a neotropical green city: an up-to-date review of the avifauna of the city of Xalapa with additional unpublished records. Urban Ecosyst 17:991–1012

González-García F, Straub R, Lobato-García JA et al (2016) Nuevos registros y notas adicionales comentadas sobre la avifauna de la ciudad de Xalapa, Veracruz, México. Acta Zool Mex 32:253–226

González-Oreja JA, De La Fuente-Díaz-Ordaz AA, Hernández-Santín L et al (2012) Can human disturbance promote nestedness? Songbirds and noise in urban parks as a case study. Landsc Urban Plan 104:9–18

González-Urrutia M (2009) Avifauna urbana en América latina: Estudios de casos. Gest Ambient 17:55–68

Grimm NB, Faeth SH, Golubiewski NE et al (2008) Global change and the ecology of cities. Science 319:756–760

Hedblom M, Murgui E (2017) Urban bird research in a global perspective. In: Murgui E, Hedblom M (eds) Ecology and conservation of birds in urban environments. Springer, New York, pp 3–10

Hernández-Lara C, González-García F, Santiago-Alarcon D (2017) Spatial and seasonal variation of avian malaria infections at places with different land uses within a Neotropical montane forest matrix. Landsc Urban Plan 157:151–160

Jetz W, Thimas GH, Joy JB et al (2012) The global diversity of birds in space and time. Nature 491:444–448

Kading RC, Reiche ASG, Morales-Betoulle ME et al (2013) Host selection of potential West Nile virus vectors in Puerto Barrios, Guatemala, 2007. Am J Trop Med Hyg 88:108–115

Kennedy C, Pincetl S, Bunje P (2011) The study of urban metabolism and its applications to urban planning and design. Environ Pollut 159:1965–1973

Komar N, Bessoff K, Diaz A et al (2012) Avian hosts of West Nile virus in Puerto Rico. Vector-Borne Zoonotic Dis 12:47–54

Lessi BF, Rodrigues-Pires JS, Batisteli AF et al (2016) Vegetation, urbanization, and bird richness in a Brazilian peri-urban area. Ornitol Neotrop 27:203–210

Leveau LM (2013) Relaciones aves–hábitat en el sector suburbano de Mar del Plata, Argentina. Ornitol Neotrop 24:201–212

Leveau LM, Leveau CM (2011) Nectarivorous feeding by the Bay-winged Cowbird (*Agelaioides badius*). Stud Neotropical Fauna Environ 46:173–175

Lima MR, Simpson L, Fecchio A et al (2010) Low prevalence of haemosporidian parasites in the introduced House Sparrow (*Passer domesticus*) in Brazil. Acta Parasitol 55:297–303

Londoño-Betancourth JC (2011) Una mirada a la diversidad ornitológica de Pereira. Bol Científico Cent Mus Mus Hist Nat 15:84–103

MacGregor-Fors I (2010) How to measure the urban-wildland ecotone: redefining "peri-urban" areas. Ecol Res 25:883–887

MacGregor-Fors I (2015) 10 scientists and 10 practitioners walk into a bar...what would they talk about? How can research and knowledge generation be co-created to better support practitioners and evidence-based decision making? In: Maddox D (ed) Global roundtable. The Nature of Cities http://www.thenatureofcities.com/2015/06/30/10-scientists-and-10-practitioners-walk-into-a-barwhat-would-they-talk-about-how-can-research-and-knowledge-generation-be-co-created-to-better-support-practitioners-and-evidence-based-decisi/. Accessed 15 May 2017

MacGregor-Fors I, Avendaño-Reyes S, Bandala VM et al (2015) Multi-taxonomic diversity patterns in a neotropical green city: a rapid biological assessment. Urban Ecosyst 18:633–647

MacGregor-Fors I, Calderón-Parra R, Meléndez-Herrada A et al (2011a) Pretty, but dangerous! Records of non-native Monk Parakeets (*Myiopsitta monachus*) in Mexico. Rev Mex Biodivers 82:1053–1056

MacGregor-Fors I, Morales-Pérez L, Schondube JE (2011b) Does size really matter? Species–area relationships in human settlements. Divers Distrib 17:112–121

MacGregor-Fors I, Ortega-Álvarez R (2011) Fading from the forest: bird community shifts related to urban park site-specific and landscape traits. Urban For Urban Gree 10:239–246

MacGregor-Fors I, Schondube JE (2011) Gray vs. green urbanization: relative importance of urban features for urban bird communities. Basic Appl Ecol 12:372–381

MacGregor-Fors I, Schondube JE (2012) Urbanizing the wild: shifts in bird communities associated to small human settlements. Rev Mex Biodivers 83:477–486

Magurran AE, McGill BJ (2011) Biological diversity – Frontiers in measurement and assessment. Oxford, New York

Martínez-Morales MA, Zuria I, Chapa-Vargas L et al (2010) Current distribution and predicted geographic expansion of the Rufous-backed Robin in Mexico: a fading endemism? Divers Distrib 16:786–797

Martins-Oliveira L, Leal-Marques R, Nunes CH et al (2012) Forrageamento de *Pitangus sulphuratus* e de *Tyrannus melancholicus* (Aves: Tyrannidae) em hábitats urbanos. Biosci J 28:1038–1050

Maruyama PK, Mendes-Rodrigues C, Alves-Silva E et al (2012) Parasites in the neighbourhood: interactions of the mistletoe *Phoradendron affine* (Viscaceae) with its dispersers and hosts in urban areas of Brazil. Flora - Morphol Distrib Funct Ecol Plants 207:768–773

Marzluff JM (2001) Worldwide urbanization and its effects on birds. In: Marzluff JM, Bowman R, Donnelly R (eds) Avian ecology and conservation in an urbanizing world. Springer, New York, pp 19–47

Marzluff JM (2016) A decadal review of urban ornithology and a prospectus for the future. Ibis 159:1–13

Marzluff JM, Bowman R, Donnelly R (eds) (2001) Avian ecology and conservation in an urbanizing world. Springer, Boston

Maxwell SL, Fuller RA, Brooks TM (2016) The ravages of guns, nets and bulldozers. Nature 536:143–145

McDonnell MJ, MacGregor-Fors I (2016) The ecological future of cities. Science 352:936–938

McKinney ML (2006) Urbanization as a major cause of biotic homogenization. Biol Conserv 127:247–260

McKinney ML (2008) Effects of urbanization on species richness: a review of plants and animals. Urban Ecosyst 11:161–176

Miller JR, Hobbs RJ (2002) Conservation where people live and work. Conserv Biol 16:330–337

Mittermeier RA, Turner WR, Larsen FW et al (2011) Global biodiversity conservation: the critical role of hotspots. In: Zachos FE, Habel JC (eds) Biodiversity hotspots. Springer, Heidelberg, pp 3–22

Molina D, Torres-Guerrero J, de la Luz Avelarde-Gómez M (2012) Riqueza de aves del Área Natural Protegida estero El Salado, Puerto Vallarta, Jalisco, México. Huitzil 13:22–33

Monnerat-Nogueira D, Rangel de Freitas A, Pinheiro da Silva C et al (2005) Estudio de la avifauna y sus ectoparásitos en un fragmento de bosque Atlántico en la Ciudad del Río de Janeiro, Brasil. Boletín SAO XV:26–36

Morales-Betoulle ME, Komar N, Panella NA et al (2013) West Nile virus ecology in a tropical ecosystem in Guatemala. Am J Trop Med Hyg 88:116–126

Moreno CE, Sánchez-Rojas G, Pineda E et al (2007) Shortcuts for biodiversity evaluation: a review of terminology and recommendations for the use of target groups, bioindicators and surrogates. Int J Environ Health 1:71–86

Nava-Diaz R, Hoogesteijn AL, Erosa MD et al (2015) Comparative study of lead concentration in feathers of urban and rural passerines in Merida, Mexico. Bull Environ Contam Toxicol 95:470–474

Nolazco S (2012) Diversidad de aves silvestres y correlaciones con la cobertura vegetal en parques y jardines de la ciudad de Lima. Bol Inf UNOP 7:4–16

Novaes WG, Cintra R (2013) Factors influencing the selection of communal roost sites by the Black Vulture *Coragyps atratus* (Aves: Cathartidae) in an urban area in Central Amazon. Zool Curitiba 30:607–614

Oliveira CB, Tonin AA, Monteiro SG (2012) Parasitismo do ácaro *Ornithonyssus bursa* em humanos no Sul do Brasil. Acta Sci Vet 40:1091

Oliveira DSF, Franchin AG, Júnior OM (2013) Disponibilidade de frutos de *Michelia champaca* L.(Magnoliaceae) e seu consumo por aves na área urbana de Uberlândia, MG. Biosci J 29:2053–2065

Ortega-Álvarez R, Calderón-Parra R (2014) Hunting the unexpected: Harris's Hawks (*Parabuteo unicinctus*) preying on bats in a Neotropical megacity. Rev Bras Ornitol 22:297–299

Ortega-Álvarez R, MacGregor-Fors I (2009) Living in the big city: effects of urban land-use on bird community structure, diversity, and composition. Landsc Urban Plan 90:189–195

Ortega-Álvarez R, MacGregor-Fors I (2011a) Spreading the word: the ecology of urban birds outside the United States, Canada, and Western Europe. Auk 128:415–418

Ortega-Álvarez R, MacGregor-Fors I (2011b) Dusting-off the file: a review of knowledge on urban ornithology in Latin America. Landsc Urban Plan 101:1–10

Pérez-García J, Monsalve-Arcilla D, Márquez-Villegas C (2015) Presencia de parásitos y enterobacterias en palomas ferales (*Columba livia*) en áreas urbanas en Envigado, Colombia. Rev Fac Nac Salud Pública 33:370–376

Pineda-López R, Malagamba A (2009) Primeros registros de presencia y reproducción del mirlo dorso rufo (*Turdus rufopalliatus*) en la ciudad de Querétaro, Querétaro, México. Huitzil 10:66–70

Pineda-López R, Malagamba A (2011) Nuevos registros de aves exóticas en la ciudad de Querétaro, México. Huitzil 12:22–27

Pineda-López R, Malagamba A, Arce I et al (2013) Detección de aves exóticas en parques urbanos del centro de México. Huitzil 14:56–64

Posa MRC, Sodhi NS (2006) Effects of anthropogenic land use on forest birds and butterflies in Subic Bay, Philippines. Biol Conserv 129:193–204

Previatto D, Mizobe R, Posso S (2013) Birds as potential pollinators of the *Spathodea nilotica* (Bignoniaceae) in the urban environment. Braz J Biol 73:737–741

Puga-Caballero A, MacGregor-Fors I, Ortega-Álvarez R (2014) Birds at the urban fringe: avian community shifts in different peri-urban ecotones of a megacity. Ecol Res 29:619–628

Ralph CJ, Geupel GR, Pyle P et al (1996) Manual de métodos de campo para el monitoreo de aves terrestres. Pacific Southwest Research Station, Forest Service, U.S. Department of Agriculture, California

Ramírez-Albores JE (2013) Riqueza y diversidad de aves de un área de la Faja Volcánica Transmexicana, Tlaxcala, México. Acta Zool Mex 29:486–512

Ramírez-Cháves HE, Pérez WA, Mejía-Egas O et al (2010) Biodiversidad en el campus de la Universidad del Cauca. Popayán, Fac Cienc Agropecu 8:104–117

Rebolo N, Fiorini VD (2010) European Starling (*Sturnus vulgaris*): population density and interactions with native species in Buenos Aires urban parks. Ornitol Neotrop 21:507–518

Rebolo-Ifrán N, Carrete M, Sanz-Aguilar A et al (2015) Links between fear of humans, stress and survival support a non-random distribution of birds among urban and rural habitats. Sci Rep 5:13723

Reis E, López-Iborra GM, Pinheiro RT (2012) Changes in bird species richness through different levels of urbanization: implications for biodiversity conservation and garden design in Central Brazil. Landsc Urban Plan 107:31–42

Ríos-Chelén AA, Salaberria C, Barbosa I et al (2012) The learning advantage: bird species that learn their song show a tighter adjustment of song to noisy environments than those that do not learn. J Evol Biol 25:2171–2180

Rodrigues LC, Araujo AC (2011) The hummingbird community and their floral resources in an urban forest remnant in Brazil. Braz J Biol 71:611–622

Rodríguez-Ruiz ER, Martínez-Sánchez I, Treviño-Carreón J (2015) Nuevos registros de aberraciones cromáticas en el plumaje de dos especies de aves en zonas urbanas de Hidalgo y Tamaulipas, México. Acta Zool Mex 31:466–469

Romero IP, Codesido M, Bilenca DN (2015) Nest building by Monk Parakeets *Myiopsitta monachus* in urban parks in Buenos Aires, Argentina: are tree species used randomly? Ardeola 62:323–333

Ruelas Inzunza E, Aguilar Rodríguez SH (2010) La avifauna urbana del Parque Ecológico Macuiltépetl en Xalapa, Veracruz, México. Ornitol Neotrop 21:87–103

Sacco AG, Bergmann FB, Ruim AM (2013) Bird assemblages in the urban area in the city of Pelotas, Rio Grande do Sul, Brazil. Biota Neotrop 13:153–162

Sainz-Borgo C (2015) Estudio del ensamblaje de aves de un parche de bosque urbano en la ciudad de Caracas, Venezuela. Acta Biol Venez 35:47–60

Santos AA, Ragusa-Netto J (2013) Toco-toucan (*Ramphastos toco*) feeding habits at an urban area in Central Brazil. Ornitol Neotrop 24:1–13

Santos AA, Ragusa-Netto J (2014) Plant food resources exploited by Blue-and-Yellow Macaws (*Ara ararauna*, Linnaeus 1758) at an urban area in Central Brazil. Braz J Biol 74:429–437

Sanz V, Caula S (2015) Assessing bird assemblages along an urban gradient in a Caribbean island (Margarita, Venezuela). Urban Ecosyst 18:729–746

Sanz V, Oviol L, Medina A et al (2010) Avifauna del estado Nueva Esparta (Venezuela): Recuento histórico y lista actual con nuevos registros de especies y reproducción. Interciencia 35:329–339

Schulze CH, Waltert M, Kessler PJ et al (2004) Biodiversity indicator groups of tropical land-use systems: comparing plants, birds, and insects. Ecol Appl 14:1321–1333

Silva CP, García CE, Estay SA et al (2015) Bird richness and abundance in response to urban form in a Latin American city: Valdivia, Chile as a case study. PLoS One 10:e0138120

Silva VL, Nicoli JR, Nascimento TC et al (2009) Diarrheagenic *Escherichia coli* strains recovered from urban pigeons (*Columba livia*) in Brazil and their antimicrobial susceptibility patterns. Curr Microbiol 59:302–308

Strewe R, Villa-De León C, Alzate J et al (2009) Las aves del campus de la Universidad del Magdalena, Santa Marta, Colombia. INTROPICA 4:79–91

Suárez-Rodríguez M, López-Rull I, Macías Garcia C (2012) Incorporation of cigarette butts into nests reduces nest ectoparasite load in urban birds: new ingredients for an old recipe? Biol Lett 9:20120931

Suárez-Rodríguez M, Montero-Montoya RD, Macías Garcia C (2017) Anthropogenic nest materials may increase breeding costs for urban birds. Front Ecol Evol 5:4

Tella JL, Canale A, Carrete M et al (2014) Anthropogenic nesting sites allow urban breeding in Burrowing Parrots *Cyanoliseus patagonus*. Ardeola 61:311–321

Teta P, Hercolini C, Cueto G (2012) Variation in the diet of Western Barn Owls (*Tyto alba*) along an urban-rural gradient. Wilson J Ornithol 124:589–596

Toledo MCB, Donatelli RJ (2010) Spectral analysis of flowers used by nectar-feeding birds in an urban area in southeastern Brazil. Braz J Biol 70:729–735

Toledo MCB, Donatelli RJ, Batista GT (2012) Relation between green spaces and bird community structure in an urban area in Southeast Brazil. Urban Ecosyst 15:111–131

United Nations (2015) World Population Prospects: The 2015 Revision. UN Department of Economic and Social Affairs, Population Division

Vargas GD, Albano AP, Fischer G et al (2011) Avian pox virus infection in a Common Barn Owl (*Tyto alba*) in southern Brazil. Pesqui Vet Bras 31:620–622

Vasconcelos MF, Mazzoni LG, Perillo A et al (2013) Long-term avifaunal survey in an urban ecosystem from southeastern Brazil, with comments on range extensions, new and disappearing species. Papéis Avulsos Zool São Paulo 53:327–344
Venter O, Sanderson EW, Magrach A et al (2016) Sixteen years of change in the global terrestrial human footprint and implications for biodiversity conservation. Nat Commun 7:12558
Villavicencio CC (2014) Aproximación a la selección de sitios de nidificación de la Gaviota Dominicana (*Larus dominicanus*, Lichtenstein 1823) en un área urbana de la región de Coquimbo (Chile) y un nuevo sustrato de nidificación. Biologist 12:33–44
Villegas M, Garitano-Zavala Á (2010) Bird community responses to different urban conditions in La Paz, Bolivia. Urban Ecosyst 13:375–391
Volpe NL, Aramburú RM (2011) Preferencias de nidificación de la cotorra Argentina (*Myiopsitta monachus*) en un área urbana de Argentina. Ornitol Neotrop 22:111–119

Chapter 3
Who Is Who in the City? Bird Species Richness and Composition in Urban Latin America

Ian MacGregor-Fors and Michelle García-Arroyo

Abstract Urban ecology in Latin America still lags behind temperate developed countries; yet, knowledge in the region has increased remarkably in the last decade. Based on previously published literature, including major global and regional reviews, we underline the prominent patterns of bird species richness and composition recorded in urban Latin America, summarizing the main factors influencing such patterns and contrasting them with those found in other regions of the world. To obtain an updated set of publications focused on bird species richness and composition in Latin America, we performed a search in the Web of Science for literature published up to 2015. After filtering out publications not considering bird species richness and/or composition, a total of 464 remained. We classified the assessed publications in relation to the study area (region and country), publication year, general aims, and main findings. We found notable similarities and differences between patterns described in studies from across the globe and from those described in previous literature reviews focused on Latin American studies. Some of the most recent publications support previously identified patterns, providing a more comprehensive understanding of how birds are responding to urbanization and its associated processes. Finally, we highlight some areas of research opportunity that could broaden our comprehension of bird ecology in relation to urbanization in Latin America. In sum, we recommend further studies assessing bird species richness and composition in urban Latin America to consider more comprehensive and comparative approaches that account for the variability of conditions occurring within urban areas, including highly developed and depauperated sites at different geographical and temporal scales.

Keywords Biodiversity statistics • Bioindicators • Bird ecology • Urban matrix • Urbanization

I. MacGregor-Fors (✉) • M. García-Arroyo
Red de Ambiente y Sustentabilidad, Instituto de Ecología, A.C.,
Carretera antigua a Coatepec 351, El Haya, Xalapa 91070, Veracruz, Mexico
e-mail: ian.macgregor@inecol.mx

I. MacGregor-Fors, J.F. Escobar-Ibáñez (eds.), *Avian Ecology in Latin American Cityscapes*, DOI 10.1007/978-3-319-63475-3_3

3.1 Introduction

The number of species present in a given area, their identity, and the distribution of their abundances have been used as fundamental ecological metrics for many decades (Jaccard 1912; MacArthur and MacArthur 1961; Gotelli and Colwell 2001; Magurran 2005). In this chapter, we focus on: (i) highlighting species richness and composition patterns that have been reported in urban areas across the globe during the past four decades; (ii) summarizing the main factors influencing these globally recognized patterns; (iii) conducting a comprehensive review of studies performed in urban areas of Latin America; (iv) contrasting these publications with those carried out in other regions of the world; and (v) highlighting areas of opportunity for future studies focusing on bird species richness and composition in Latin America. It is noteworthy that this chapter is devoted exclusively to patterns of species richness and composition (see Chap. 4 for information related to avian abundances and demographic dynamics in urban areas). For practicality, we used the 'community' concept to describe variations of avian species richness and composition in urban areas, following the simple and flexible definition provided by Fauth et al. (1996): "collection of species [in this case birds] occurring in the same space and time". We also considered publications that targeted focal bird species or groups (e.g., based on their phylogeny and/or use of resources) for our species composition assessments (namely 'assemblages', 'ensembles', and 'guilds'; sensu Fauth et al. 1996).

3.2 Global Species Richness and Composition Patterns

Previous literature reviews, starting with the keystone contributions of Marzluff et al. (2001), Chace and Walsh (2006), and Evans et al. (2009), have identified several general patterns describing bird species richness and composition in urban areas across the globe. It is noteworthy that each of these three leading literature reviews recognizes a clear underrepresentation of studies carried out in tropical urban settings, where biodiversity peaks but urbanization is expected to increase (Myers et al. 2000; United Nations 2015; see Chap. 1 for a detailed discussion). Fortunately, studies performed in tropical and subtropical cities have risen in recent years, showing a marked increase since 2003 (González-Urrutia 2009; Ortega-Álvarez and MacGregor-Fors 2011a; Chap. 2).

3.2.1 *Species Richness*

Previous literature reviews have found several patterns regarding the number of bird species that dwell in a given urban location, many of them with interesting exceptions. One of the most generalizable patterns is that urbanization tends to negatively

Fig. 3.1 Urban-dwelling birds using human-made structures in the city of Xalapa (Veracruz, Mexico). *Left*: Ferruginous Pygmy-Owl (*Glaucidium brasilianum*) perched on a tangle of electric and telephone wires. *Right*: Acorn Woodpeckers (*Melanerpes formicivorus*) perched and foraging at the top of a wooden telephone pole (Photos: Juan F. Escobar-Ibáñez)

affect bird species richness (Marzluff et al. 2001; Chace and Walsh 2006); yet, it has also been shown to peak at intermediate levels of urbanization (reviewed by Evans et al. 2009; also see Blair 1996; Lepczyk et al. 2008), rise with urbanization (Aldrich and Coffin 1980), and even show no differences between urban and nonurban sites (Sodhi 1992). These patterns are given by a diverse array of variables that can shape the results of studies addressing avian richness in urbanization gradients, as well as in urban–nonurban comparisons, of which temporality (Poague et al. 2000) and other social and economic predictors head the list (e.g., history, size, location, urbanization intensity, spatial heterogeneity; Edgar and Kershaw 1994; Germaine et al. 1998; Blair 2001).

As would be expected, urban vegetation heavily influences the number of bird species that can dwell within an urban area. In fact, most studies focused on urban bird species richness have concentrated their efforts within greenspaces (Chace and Walsh 2006; Evans et al. 2009; Malagamba-Rubio et al. 2013). Several greenspace characteristics have been identified to shape bird species richness in urban areas, showing two general patterns: (i) increases with vegetation complexity and diversity (Emlen 1974; Beissinger and Osborne 1982; Schwarzenberger and Dean 2003; Donnelly and Marzluff 2006; MacGregor-Fors 2008); and (ii) increases with patch size (Gavareski 1976; Mills et al. 1989; Chamberlain et al. 2007). 'Gray' urban infrastructure (DeGraff and Wentworth 1986; Sacchi et al. 2002) has also shown to determine bird abundances, essentially because some species are able to exploit urban conditions and resources (see Chaps. 4 and 5; Fig. 3.1). Other urban-related variables have been shown to correlate with bird species richness, including supplementary resources (e.g., nest boxes, bird feeders, urban litter; Gaston et al. 2005; Glue 2006; Fuller et al. 2008), human activity (e.g., passing pedestrians, vehicle traffic; Knight and Gutzwiller 1995; Miller et al. 1998), collisions (e.g., window, power-line, and car strikes; Klem 1989; Codoner 1995; Chap. 7), domestic and urban-related predators (e.g., dogs, cats, corvids; Kristan et al. 2003; Jokimäki et al. 2005), competition with exotic/invasive bird species (MacGregor-Fors et al. 2010), and disease transmission (e.g., trichomoniasis; Boal and Mannan 1999).

3.2.2 *Species Composition*

Regarding the identity of birds in urban areas, several approaches have been used, including the assessment of changes in the species composition of communities, assemblages, ensembles, and guilds (sensu Fauth et al. 1996). A global literature review suggested that omnivore, granivore, and cavity-nesting species benefit from urbanization (Chace and Walsh 2006). However, a subsequent regional review, focused on urban birds from tropical and subtropical regions highlighted that, in addition to omnivores and granivores, other feeding guilds such as nectarivores, frugivores, and insectivores comprise an important proportion of the avifauna of cities in Australia, Singapore, and Mexico and Brazil, respectively (Ortega-Álvarez and MacGregor-Fors 2011a). Along with guilds and functional (often trophic) groups, other classifications have been used to assess the response of bird communities to urbanization. For example, studies have shown urbanization to have a greater negative effect on natives than exotics, as well as residents over migrants (Allen and O'Connor 2000; Poague et al. 2000), and specialists over generalists (Lancaster and Rees 1979; Johnston 2001; Fraterrigo and Wiens 2005; Smith and Wachob 2006).

3.2.3 *Updated Global Literature Review*

To thoroughly review and update the existing literature related to bird species richness and composition in urban areas around the globe, we used the Web of Science (http://webofknowledge.com) platform, which has recently broadened its databases to include additional journals, conference proceedings, book citation indices, and regional databases (e.g., SciELO, Russian Science Citation Index). We constrained our search for publications dating from 1900 to 2015, using the following databases: Web of Science Core Collection, Biological Abstracts, Current Contents Connect, Derwent Innovations Index, FSTA, KCI-Korean Journal Database, Russian Science Citation Index, SciELO Citation Index, and Zoological Record. To retrieve a broad initial sample of publications, we used the following keyword combination: ('urban' OR 'city') AND ('bird' OR 'avian') AND ('species richness' OR 'richness' OR 'composition' OR 'turnover' OR 'diversity'). We considered 'diversity' as a keyword for this chapter, as many researchers working with this topic base their studies within a diversity framework; however, we filtered publications that focused specifically on species richness and/or composition findings.

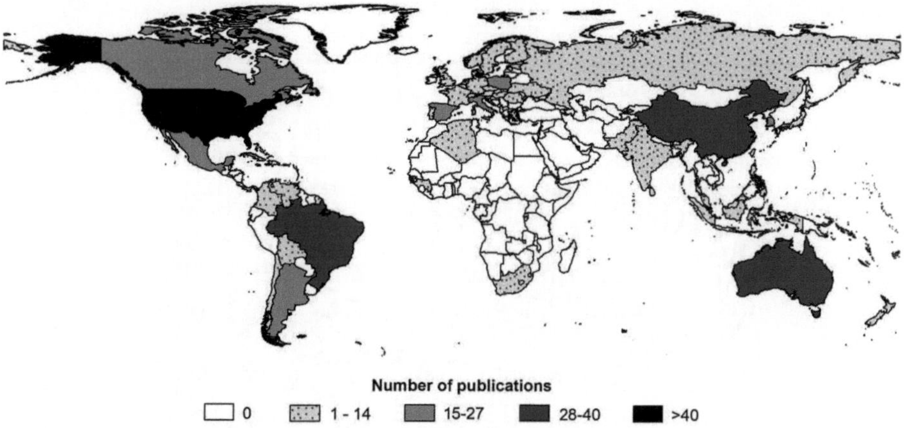

Fig. 3.2 Worldwide geographical representation of the number of reviewed publications by country

Our general search resulted in a total of 1434 publications, starting with studies from the late 1970s performed in Finland, North America, and Russia (e.g., Huhtalo and Järvinen 1977; Korolkova 1978; Lancaster and Rees 1979; Fig. 3.2). We performed a thorough screening of publication titles and abstracts to filter out those that fell outside of the scope of this chapter, which resulted in a total of 464 retrieved publications. We classified these publications by country, year of publication, general aims, and main findings (see Fig. 3.3 for a word cloud showing the most frequent words in the abstracts).

Our analysis of the 464 retrieved publications showed that most studies were carried out along urbanization gradients (which often, but not necessarily, include extra-urban systems; sensu MacGregor-Fors et al. 2010; Lancaster and Rees 1979; Rolando et al. 1997; van Rensburg et al. 2009; Verma and Murmu 2015), followed by those set up at specific urban locations (e.g., parks, gardens, preserves; Biaduń 1994; Day 1995; Donnelly and Marzluff 2004; Yang et al. 2015a), and considering urban versus nonurban comparisons (e.g., urban vs. 'rural', 'suburban', and/or 'natural' locations; Holtz 1998; Sorace 2001; Soh et al. 2006; Kopij 2011). It is noteworthy that some of these studies did not compare their results with adjacent systems (Luniak et al. 1986; Jones and Wieneke 2000; Dassanayaka and Mahaulpatha 2009; Strohbach et al. 2014). Among the most best-studied intra-urban conditions across the globe were: (i) parks (e.g., Tomiałojć 1998; Morneau et al. 1999; Densmore and French 2005; Zhang and Miao 2013); (ii) residential gardens (e.g., Catterall 2004; Chamberlain et al. 2004; Lerman et al. 2012; Paker et al. 2014); (iii) urban preserves (e.g., Mörtberg 2001; Drinnan 2005; Chapman and Reich 2007; Lin et al. 2013); (iv) wetlands (e.g., Rosa et al. 2003; Smith and Wachob 2006; Bensizerara et al. 2013; Sun et al. 2015); and (v) university campuses (e.g., Takeuchi et al. 2010; Shultz et al. 2012; Gatesire et al. 2014). Regarding studies that included nonurban systems, most comparisons of urban bird communities were made with those of native systems (e.g., woodland, shrubland, grassland; Yeoman and Nally 2005;

Fig. 3.3 Word cloud showing the most frequent words ($n = 250$; after eliminating articles, prepositions, conjunctions, interjections, and other common words) used in the abstracts of the 464 analyzed publications performed worldwide

Santos et al. 2008; Caula et al. 2010), followed by other human-modified systems (e.g., suburban, farmland, pastures; Parsons et al. 2003; Filippi-Codaccioni et al. 2009; Le Roux et al. 2015).

From our updated global review, we identified several general patterns regarding the species richness and composition of bird communities in urban areas that are consistent with those underlined in previous reviews of worldwide publications (Marzluff et al. 2001; Chace and Walsh 2006; Evans et al. 2009; Ortega-Álvarez and MacGregor-Fors 2011a): (i) bird species richness decreases with urbanization (e.g., Vallejo et al. 2009; Ciach 2012; Barth et al. 2015; Schneider et al. 2015); (ii) bird species richness increases with urban vegetation complexity and diversity (e.g., Melles et al. 2003; White et al. 2005; Husté et al. 2006; Kang et al. 2015), as well as with the age of the urban area (e.g., Jones 1981; Mason 1985; Fernández-Juricic 2000; Filippi-Codaccioni et al. 2008) and greenspace patch size (e.g., Fernández-Juricic 2000; Donnelly and Marzluff 2004; Palmer et al. 2008; Suk et al. 2014); (iii) species richness decreases with human activity (e.g., pedestrian and car traffic; Fernández-Juricic 2001; Platt and Lill 2006; Young et al. 2007; Verma and Murmu

2015); (iv) species identified to thrive in cities are commonly generalists (e.g., Conole and Kirkpatrick 2011; Davis and Wilcox 2013; Vignoli et al. 2013) and/or exotics (e.g., van Heezik et al. 2008; Loss et al. 2009; Gagné and Fahrig 2011; Luck et al. 2013), with omnivorous, granivorous, and insectivorous birds heading the list (i.e., omnivores: Clergeau et al. 1998; Smith 2003; Walker and Shochat 2010; Huang et al. 2015; granivores: Fox and Hockey 2007; Blair and Johnson 2008; Sengupta et al. 2014; insectivores: Zhou and Chu 2014; Kopij 2015).

Although some of these patterns have been considered in previous literature reviews (Marzluff et al. 2001; Chace and Walsh 2006; Evans et al. 2009; Ortega-Álvarez and MacGregor-Fors 2011a), the results of our updated global review identified some undermentioned new patterns, some of which are contradictory with generalized ones, such as: (i) bird species richness increases with urbanization across the globe (e.g., Pautasso and Dinetti 2009; Kopij 2011; Palita et al. 2011; Trammell et al. 2011); (ii) species richness increases with tree and shrub cover (e.g., Luther et al. 2008; Chong et al. 2014; Ikin et al. 2014; Yang et al. 2015b), native vegetation (e.g., Pennington et al. 2008; Burghardt et al. 2009; Davis and Wilcox 2013; Ikin et al. 2013), and the presence of water bodies (e.g., Melles et al. 2003; Smith 2003; Chamberlain et al. 2007; Shwartz et al. 2008); (iii) species richness decreases with building density (e.g., Maeda 1998; Palomino and Carrascal 2005; Caula et al. 2010; Schneider and Miller 2014), greenspace isolation (e.g., Murgui 2007; Lee et al. 2010; Smith and Chow-Fraser 2010; Shanahan et al. 2011), and percentage of impervious cover (e.g., Luck et al. 2013; Song 2015); and (iv) human commensal species are benefited by urbanization (e.g., Ciach 2012; Wood et al. 2015).

3.3 Species Richness and Composition in Urban Latin America

In 2011, Ortega-Álvarez and MacGregor-Fors (2011b; hereafter referred to as 'previous regional literature review') reviewed 84 studies published from 1974 to 2009 that focused on birds from urban Latin America. Studies assessing avian species richness and/or composition included comparative studies performed on setups involving urban–nonurban dichotomies (Fontana 2005; Tampson and Petry 2008), urban gradients (Reynaud and Thioulouse 2000; Leveau and Leveau 2004; Faggi and Perepelizin 2006; Faggi et al. 2006), intra-urban sites (e.g., residential areas: Argel-de-Oliveira 1995; Soares 2004; urban parks: Scherer et al. 2005; Gómez 2006; urban gardens: de Lima and Aleixo 2000; university campuses: MacGregor-Fors 2005; Marín-Gómez 2005), and nonurbanized systems within urban areas (e.g., beaches: Acosta-Ramos and Batista-Daudt 2005; da Silva 2006; lakes: Silva and Blamires 2007; secondary vegetation: Rivera-Gutiérrez 2006; da Silva 2006; urban forest remnants: Monnerat-Nogueira et al. 2005). In terms of the amount of information, previous comparisons have shown that research focused on bird diversity in urban Latin America lags behind that of other regions, mainly those from temperate developed countries (Ortega-Álvarez and MacGregor-Fors 2011b; MacGregor-Fors and Ortega-Álvarez 2013; also see Chaps. 1 and 2).

3.4 Updated Literature Review of Urban Latin America

To review the existing knowledge of bird species richness and composition from Latin American urban settings, we used the filtered subset of 464 publications and grouped together all studies that were performed in any of the following 47 Latin American countries: Anguilla, Antigua and Barbuda, Argentina, Aruba, Bahamas, Barbados, Belize, Bermuda, Bolivia, Brazil, British Virgin Islands, Cayman Islands, Chile, Colombia, Costa Rica, Cuba, Dominica, Dominican Republic, Ecuador, El Salvador, Grenada, Guadeloupe, Guatemala, Guiana, French Guiana, Haiti, Honduras, Jamaica, Martinique, Mexico, Montserrat, Netherlands Antilles, Nicaragua, Panama, Paraguay, Peru, Puerto Rico, Saint Barthélemy, Saint Kitts and Nevis, Saint Lucia, Saint Martin, Saint Vincent and the Grenadines, Suriname, Trinidad and Tobago, Turks and Caicos Islands, Uruguay, and Venezuela.

Of the 464 publications, 92 of them corresponded to studies that were both performed in Latin American urban areas and focused on urban bird species richness and composition, representing ~20% of the total retrieved publications (Fig. 3.4). This represents a noticeable increase in publications focused on urban bird species richness and composition in Latin America when compared with the previous regional literature review, with publications focused on these approaches from 2009 to 2015 (n = 50) representing a similar number of studied than those found to address ecological questions in the previous regional review.

When comparing the 92 reviewed publications from urban Latin America with those reviewed in the previous regional literature review, we found both similarities and differences, with the recent publications (2009–2015) strengthening the evidence for the patterns recognized across the world and shedding light on a more comprehensive understanding of the way in which birds respond to urbanization in Latin America. We identified several patterns that agree with those reported in the previous global and regional literature reviews: (i) species richness increases with vegetation structure and complexity (e.g., Cardozo et al. 2008; Carbó-Ramírez and Zuria 2011; Toledo et al. 2012; Leveau 2013; Domínguez-López and Ortega-Álvarez 2014), greenspace patch size (e.g., Garitano-Zavala and Gismondi 2003; Suarez-Rubio and Thomlinson 2009; dos Santos and Cademartori 2010; Carbó-Ramírez and Zuria 2011), and environmental heterogeneity (e.g., Manhães and Loures-Ribeiro 2005; Silva and Blamires 2007; Navarro et al. 2011; González-Oreja et al. 2012a, b); (ii) species richness decreases with human activity (e.g., pedestrians, vehicle traffic; Heil et al. 2007; Cruz and Piratelli 2011; MacGregor-Fors and Schondube 2011; González-Oreja et al. 2012a; MacGregor-Fors et al. 2012) and urban infrastructure (e.g., buildings, streets, squares; Carbó-Ramírez and Zuria 2011; MacGregor-Fors and Schondube 2011; MacGregor-Fors et al. 2012;

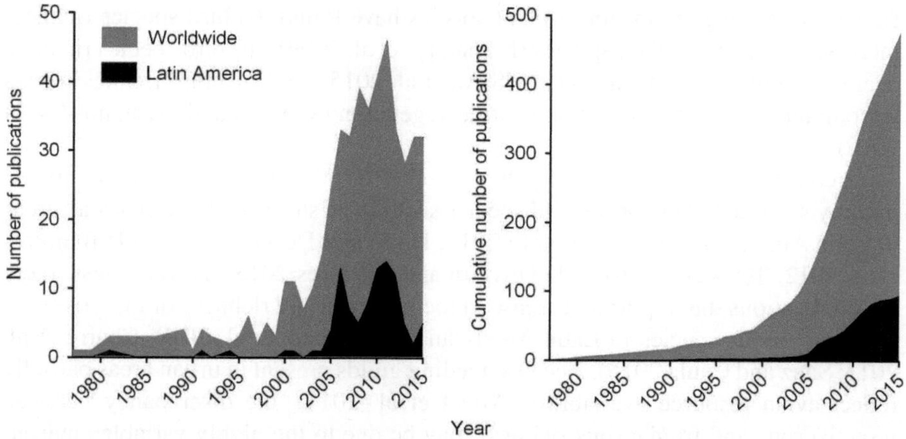

Fig. 3.4 Number of publications focused on avian species richness and composition in urban Latin America and across the globe through time

Sacco et al. 2015; Silva et al. 2015); and (iii) insectivorous birds are successful in urban areas (Germain et al. 2008; Fernández et al. 2009; Maragliano et al. 2009; Bispo and Scherer-Neto 2010; Brummelhaus et al. 2012), as well as generalist and exotic species (Garaffa et al. 2009; Ortega-Álvarez and MacGregor-Fors 2009; Presti and Echevarria 2009; Athiê and Dias 2010; Fontana et al. 2011). As pinpointed in the global literature reviews, recent studies developed in Latin America have been carried out in specific urban locations (e.g., parks, university campuses; Lopes and dos Anjos 2006; Muñoz et al. 2007; MacGregor-Fors and Ortega-Álvarez 2011; González-Oreja et al. 2012b; Charre et al. 2013), along urbanization gradients (Garaffa et al. 2009; Juri and Chani 2009; Villegas and Garitano-Zavala 2010; Leveau et al. 2015), and comparing urban versus nonurban sites (e.g., agriculture, native ecosystems; Rodríguez-Estrella 2007; Kennedy et al. 2010; Dario and De Vincenzo 2011; MacGregor-Fors et al. 2012; Domínguez-López and Ortega-Álvarez 2014). Multiple Latin American studies have also included water bodies and wetlands either within or near urban locations, underlining their importance as habitats for many bird species (Faggi et al. 2006; Silva and Blamires 2007; Molina et al. 2012; Rosselli and Stiles 2012). It is noteworthy that Latin American cities have been shown to shelter high avian diversity, mainly within their greenspace networks (Escobar-Ibáñez and MacGregor-Fors 2016). For instance, over 340 species have been recorded within the boundaries of the city of Xalapa, a small-sized Mexican city embedded within a cloud–forest matrix (González-García et al. 2014, 2016).

In addition to well-documented patterns of urban bird richness and composition in Latin America, we found interesting information that adds to our comprehension of how urbanization affects birds outside highly studied regions. For instance, we found that some recent studies in Latin America have started to consider a wider array of variables that may be acting as drivers of urban bird species richness values, such as assessing the origin of urban plant species (i.e., native/exotic). Results from studies addressing the origin of plant species have found: (i) bird species richness increases with native plant species (Fernández et al. 2009); (ii) bird species richness decreases with exotic plant species (Silva et al. 2015); and (iii) bird richness peaks in combination of both native and exotic vegetation (Villegas and Garitano-Zavala 2010).

In contrast with composition patterns found in other regions across the globe, insectivorous and frugivorous bird species seem to be successful within urban areas in Latin America (Cruz and Piratelli 2011; Dario and De Vincenzo 2011; Blamires et al. 2012; Teles et al. 2012; de Oliveira and Blamires 2013). Nevertheless, there are publications that report a decrease in the presence and richness of insectivorous and frugivorous species in Latin American cities (Franz et al. 2010; Charre et al. 2013; Sanz and Caula 2015). Because feeding guilds present in urban areas partially reflect avian resource availability (Vogel et al. 2011), the discrepancy between insectivorous and frugivorous richness may be due to the highly variable environmental and structural heterogeneity for Latin American cities, as well as the species pool present in the surrounding systems. Altogether, species richness loss and composition shifts related to avian traits in urban systems (e.g., Suarez-Rubio and Thomlinson 2009; Franz et al. 2010; González-Oreja 2011) suggest an urban 'semi-permeability' process for birds (MacGregor-Fors 2010), although some urban 'avoider' species often remain within large urban greenspaces (Fig. 3.5), as suggested by Fischer et al. (2015).

3.5 Future Directions

Diversity measurements of bioindicator groups, such as birds, can improve our understanding of how wildlife species respond to human disturbances (Moreno et al. 2007). Yet, measuring some of the emergent properties of communities, as the ones addressed in this chapter (i.e., species richness, composition), can be technically challenging. As field and analytical procedures are continuously changing and evolving with our need to understand biodiversity, it is often complicated to contrast results between the wide array of metrics, which are often used indifferently. Thus, we suggest researchers to use up-to-date and robust methods to analyze their data, considering recommendations included by Magurran and McGill (2011) as well as more recent ones. For instance, when contrasting species richness among environmental conditions, we suggest the use of rarefied species richness values (Gotelli and Colwell 2001), extrapolated if needed, followed by the contrasting of confidence intervals to establish inferential results (MacGregor-Fors and Payton 2013). Regarding procedures to assess the composition of urban avian communities, it is crucial to appropriately select the level of analysis (e.g., community, assemblage, trophic groups), the approach to assess β–diversity (e.g., differentiation, scaling, turnover, nestedness), and the metric to quantify it (Koleff et al. 2003; Moreno and Rodríguez 2010; Baselga and Leprieur 2015).

Fig. 3.5 Colorful and/or charismatic birds that can be easily recorded in Latin American urban greenspaces. *Left*: Bananaquit (*Coereba flaveola*), Parque Natural Chico Mendes, Sorocaba, Brasil (Photo: IM-F). *Center*: Lineated Woodpecker (*Dryocopus lineatus*) Parque Fundadores, Armenia, Quindío. *Right*: Rufous-tailed Hummingbird (*Amazilia tzacatl*), Campus Universidad del Quindio, Armenia, Quindío (Photos: Oscar H. Marín Gómez)

As has been identified in the regional literature review, there is a clear underrepresentation of Latin American countries contributing to the urban bird knowledge. Additionally, research focuses in large and iconic cities (see Chap. 2), leaving aside the heterogeneity of urban scenarios and the associated response of their related avifaunas (see Chap. 9 for an in-depth analysis). Taking into account the information provided in previous regional literature reviews and our updated one, we identified some areas of opportunity that could aid in broadening the scope of potential factors that drive bird species richness and composition in urban Latin America.

The complex nature of urbanized landscapes raises an important need for researchers who are assessing urban bird species richness and composition to better incorporate contextual components and scale-dependent ecological variables into their studies (Rodríguez-Estrella 2007; Walker and Shochat 2010; Trammell et al. 2011; Lerman et al. 2012; Molina et al. 2012). As noted previously, many studies developed in urban Latin America have focused their surveys on single and specific types of sites, such as greenspaces, with little information regarding the context in which study areas are embedded. In this sense, we have identified three important considerations for establishing such a context. First, most studies ignore the urban matrix that surrounds target urban greenspaces. If we aim to understand how urban systems work and how they relate to biodiversity, it is crucial to assess highly urbanized systems to get a broader ecological picture (Fuller et al. 2009; Nilon et al. 2011; Sushinsky et al. 2013; Escobar-Ibáñez and MacGregor-Fors 2016). Thus, we suggest researchers to follow comparative approaches, such as using highly developed versus lowly developed urban sites, urbanization intensity gradients (McDonnell and Pickett 1990; Hahs and McDonnell 2006; Ortega-Álvarez and MacGregor-Fors 2009; Leveau et al. 2015; Silva et al. 2015), and citywide surveys that include representative samples of the physical, environmental, and socioeconomic variances that occur within cities (Turner 2003).

Second, broadening the geographical scale of studies could provide, in general, the context of the landscape of which any given city is a part of (La Sorte et al. 2014; Lee and Carroll 2014; Tryjanowski et al. 2015). Depending on the level of analysis and scope, this could also add to our knowledge on more regional patterns that better reflect the variability of urban conditions and scenarios, as has recently been accomplished at a global scale (Aronson et al. 2014; Sol et al. 2014). Although intensive studies of focal cities are of major importance, comparative studies at different scales will undoubtedly broaden our comprehension of the ways in which avian communities are responding and adapting to urbanization in Latin America (Chap. 6). Another aspect to consider is the temporal scale. The time span of most ecological bird studies in urban areas is relatively short (Marzluff et al. 2001), which also applies to those performed in Latin America (pers. obs.). While short-term studies are helpful in many circumstances (MacGregor-Fors et al. 2015), mid-term and long-term studies are essential to understand the temporal dynamics related to the responses of bird communities to urbanization (Ormond et al. 2014; Strohbach et al. 2014; Escobar-Ibáñez and MacGregor-Fors 2016). For instance, temporal analyses of avian communities can provide information on the differential use of resources, which in fact can drive both urban bird species richness and composition (Catterall et al. 2010; Shultz et al. 2012; see Chaps. 5 and 6).

Third, there has been growing interest toward studying how biodiversity responds to land-use change at the urban fringe, where cities tend to sprawl. Given that urbanization implies the modification or replacement of preexisting systems, including changes in the terrain, hydrology, and vegetation, among other components, their boundaries can act as filters for wildlife species (Croci et al. 2008). The area where the urban core (intra-urban area) merges with adjacent systems, namely the 'ecotones' in which urban areas interact with immediate nonurban systems (peri-urban areas), have been proven to be ecologically relevant for bird communities, with richer communities found in peri-urban areas when compared with intra-urban sites with similar traits (MacGregor-Fors 2010; Puga-Caballero et al. 2014). These kinds of studies focused on bird species richness and composition are quite scarce in the overall literature, but there is a good representation of Latin American cities among them (Garaffa et al. 2009; MacGregor-Fors 2010; Puga-Caballero et al. 2014). Although some studies have begun to focus on the conceptual framework of 'peri-urban' areas (e.g., Iaquinta and Drescher 2000; Tjallingii 2000; Snep et al. 2006), this topic is incipient and could provide an important ecological foundation for developing and implementing practices that address issues presented by urban sprawl.

Finally, there is a topic that requires special attention to understand how urbanization can shape regional avifaunas. Previous studies performed across the globe have suggested that biotic homogenization takes place in urban areas (McKinney 2006; Devictor et al. 2007; Ortega-Álvarez and MacGregor-Fors 2009; Murthy et al. 2016); yet, recent global-scale studies show contrastingly different results regarding this phenomenon. On the one hand, Aronson et al. (2014) showed that, using an incidence-based worldwide urban bird dataset considering 54 urban areas, cities retain similar composition patterns within biogeographical realms, suggesting that urban avifaunas have not been taxonomically homogenized at the global scale. On the other hand, Sol et al. (2014), using an abundance-based dataset of 22 cities, show that species are being filtered by urbanization due to their lack of capability in adapting to the novel conditions, mainly related to the use and exploitation of urban resources and avoidance of urban-related risks (see Chap. 7), reducing β-diversity, and ultimately promoting biotic homogenization. Thus, there is still a dearth of knowledge on the scales and factors related to potential homogenization of biota driven by urbanization, making this a highly important topic to tackle. However, researchers need to be careful when assessing biotic homogenization, as some fundamental concepts are prone to being misconstrued (Olden 2006; see Chap. 9). Lastly, we strongly encourage that all studies, regardless of their aims, scopes, and/or scales should devote a section that provides feasible urban management and planning strategies founded on their evidence-based findings.

Acknowledgments The authors are greatly thankful to Jeffrey G.-H. Lee and Rafael Rueda Hernández for their helpful comments that enhanced the quality and clarity of the chapter, as well as Carlos Chávez Zichinelli, Miguel Ángel Gómez Martínez, and Julian Avila Campos for their support with figures and abstract screening. MG-A acknowledges the scholarship and financial support provided by the National Council of Science and Technology (CONACYT 416452) and the Master's Program of the Instituto de Ecología, A.C. (INECOL).

References

Acosta-Ramos L, Batista-Daudt R (2005) Avifauna urbana dos balneários de Tramandaí e Imbé, litoral norte do Rio Grande do Sul. Biotemas 18:181–191

Aldrich JW, Coffin RW (1980) Breeding bird populations from forest to suburbia after thirty-seven years. Am Birds 34:3–7

Allen AP, O'Connor RJ (2000) Hierarchical correlates of bird assemblage structure on northeastern USA lakes. Environ Monit Assess 62:15–37

Argel-de-Oliveira MM (1995) Aves e vegetação em um bairro residencial da cidade de São Pauolo (São Pauolo, Basil). Rev Bras Zool 12:81–92

Aronson MFJ, La Sorte FA, Nilon CH et al (2014) A global analysis of the impacts of urbanization on bird and plant diversity reveals key anthropogenic drivers. Proc R Soc Lond B Biol Sci 281:20133330

Athiê S, Dias MM (2010) Avian community composition in an urban park in central São Paulo state, southeastern Brazil. Rev Bras Zoociências 12:67–73

Barth JB, FitzGibbon SI, Wilson RS (2015) New urban developments that retain more remnant trees have greater bird diversity. Landsc Urban Plan 136:122–129

Baselga A, Leprieur F (2015) Comparing methods to separate components of beta diversity. Methods Ecol Evol 6:1069–1079

Beissinger SR, Osborne DR (1982) Effects of urbanization on avian community organization. Condor 84:75–83

Bensizerara D, Chenchouni H, Bachir AS et al (2013) Ecological status interactions for assessing bird diversity in relation to a heterogeneous landscape structure. Avian Biol Res 6:67–77

Biaduń W (1994) The breeding avifauna of the parks and cemeteries of Lublin (SE Poland). Acta Ornithol 29:1–13

Bispo AÂ, Scherer-Neto P (2010) Taxocenose de aves em um remanescente da Floresta com Araucária no sudeste do Paraná, Brasil. Biota Neotrop 10:121–130

Blair RB (1996) Land use and avian species diversity along an urban gradient. Ecol Appl 6:506–519

Blair RB (2001) Birds and butterflies along urban gradients in two ecoregions of the United States: is urbanization creating a homogeneous fauna? In: Lockwood J, McKinney M (eds) Biotic homogenization. Springer, New York, pp 33–56

Blair RB, Johnson EM (2008) Suburban habitats and their role for birds in the urban–rural habitat network: points of local invasion and extinction? Landsc Ecol 23:1157–1169

Blamires D, de Oliveira-Neto JB, Ponciano JG et al (2012) Aves do Clube Recreativo de Iporá, Estado de Goiás, Brasil. Rev Bras Zoociências 14:57–70

Boal CW, Mannan RW (1999) Comparative breeding ecology of Cooper's Hawks in urban and exurban areas of southeastern Arizona. J Wildl Manag 63:77–84

Brummelhaus J, Bohn MS, Petry MV (2012) Effect of urbanization on bird community in riparian environments in Caí River, Rio Grande do Sul, Brazil. Biotemas 25:81–96

Burghardt KT, Tallamy DW, Shriver WG (2009) Impact of native plants on bird and butterfly biodiversity in suburban landscapes. Conserv Biol 23:219–224

Carbó-Ramírez P, Zuria I (2011) The value of small urban greenspaces for birds in a Mexican city. Landsc Urban Plan 100:213–222

Cardozo G, Beltzer A, Collins P (2008) Variación primavero–estival de la diversidad y abundancia de la comunidad de aves en la Reserva Ecológica de la Ciudad Universitaria U.N.L. "El Pozo.". In: Aceñolaza FG (ed) Temas de la biodiversidad del litoral fluvial Argentino III. Consejo Nacional de Investigaciones Científicas y Técnicas, San Miguel de Tucumán, pp 367–387

Catterall CP (2004) Birds, garden plants and suburban bushlots: where good intentions meet unexpected outcomes. In: Lunney D, Burgin S (eds) Urban wildlife: more than meets the eye. Royal Zoological Society of New South Wales, Mosman, pp 21–31

Catterall CP, Cousin JA, Piper S et al (2010) Long-term dynamics of bird diversity in forest and suburb: decay, turnover or homogenization? Divers Distrib 16:559–570

Caula SA, Sirami C, Marty P et al (2010) Value of an urban habitat for the native Mediterranean avifauna. Urban Ecosyst 13:73–89

Chace JF, Walsh JJ (2006) Urban effects on native avifauna: a review. Landsc Urban Plan 74:46–69

Chamberlain DE, Cannon AR, Toms MP (2004) Associations of garden birds with gradients in garden habitat and local habitat. Ecography 27:589–600

Chamberlain DE, Gough S, Vaughan H et al (2007) Determinants of bird species richness in public green spaces. Bird Study 54:87–97

Chapman KA, Reich PB (2007) Land use and habitat gradients determine bird community diversity and abundance in suburban, rural and reserve landscapes of Minnesota, USA. Biol Conserv 135:527–541

Charre GM, Hurtado JAZ, Néve G et al (2013) Relationship between habitat traits and bird diversity and composition in selected urban green areas of Mexico City. Ornitol Neotrop 24:275–293

Chong KY, Teo S, Kurukulasuriya B et al (2014) Not all green is as good: different effects of the natural and cultivated components of urban vegetation on bird and butterfly diversity. Biol Conserv 171:299–309

Ciach M (2012) The winter bird community of rural areas in the proximity of cities: low density and rapid decrease in diversity. Pol J Ecol 60:193–199

Clergeau P, Savard J-PL, Mennechez G et al (1998) Bird abundance and diversity along an urban–rural gradient: a comparative study between two cities on different continents. Condor 100:413–425

Codoner NA (1995) Mortality of Connecticut birds on roads and at buildings. Conn Warbler 15:89–98

Conole LE, Kirkpatrick JB (2011) Functional and spatial differentiation of urban bird assemblages at the landscape scale. Landsc Urban Plan 100:11–23

Croci S, Butet A, Clergeau P (2008) Does urbanization filter birds on the basis of their biological traits? Condor 110:223–240
Cruz BB, Piratelli AJ (2011) Avifauna associated to an urban extend of the Sorocaba river, Southeastern Brazil. Biota Neotrop 11:255–264
da Silva RR (2006) Estrutura de uma comunidade de aves em Caxias do Sul, Rio Grande do Sul, Brasil. Biociencias 14:27–36
Dario FR, De Vincenzo MCV (2011) Avian diversity and relative abundance in a restinga forest of Sao Paulo, Brazil. Trop Ecol 52:25–33
Dassanayaka T, Mahaulpatha DWA (2009) Avifaunal species richness and diversity at the premises of the University of Sri Jayewardenepura, Nugegoda, Sri Lanka. Vidyodaya J Sci 14:41–51
Davis RA, Wilcox JA (2013) Adapting to suburbia: bird ecology on an urban–bushland interface in Perth, Western Australia. Pac Conserv Biol 19:110–120
Day TD (1995) Bird species composition and abundance in relation to native plants in urban gardens, Hamilton, New Zealand. Notornis 42:172–186
de Lima F, Aleixo A (2000) Notas sobre algumas aves em ambientes antropizados da cidade de Campinas, São Paulo, Brasil. Bol CEO 14:2–6
de Oliveira JB, Blamires D (2013) Aves do campus do Instituto Federal de Educação em Iporá, Estado de Goiás. Semina Cienc Biol Saude 34:45–54
DeGraff RM, Wentworth JM (1986) Avian guild structure and habitat associations in suburban bird communities. Urban Ecol 9:399–412
Densmore P, French K (2005) Effects of recreation areas on avian communities in coastal New South Wales parks. Ecol Manag Restor 6:182–189
Devictor V, Julliard R, Couvet D et al (2007) Functional homogenization effect of urbanization on bird communities. Conserv Biol 21:741–751
Domínguez-López ME, Ortega-Álvarez R (2014) The importance of riparian habitats for avian communities in a highly human-modified Neotropical landscape. Rev Mex Biodivers 85:1217–1227
Donnelly R, Marzluff JM (2004) Importance of reserve size and landscape context to urban bird conservation. Conserv Biol 18:733–745
Donnelly R, Marzluff JM (2006) Relative importance of habitat quantity, structure, and spatial pattern to birds in urbanizing environments. Urban Ecosyst 9:99–117
dos Santos MFB, Cademartori CV (2010) Estudo comparativo da avifauna em áreas verdes urbanas da região metropolitana de Porto Alegre, sul do Brasil. Biotemas 23:181–195
Drinnan IN (2005) The search for fragmentation thresholds in a southern Sydney suburb. Biol Conserv 124:339–349
Edgar DR, Kershaw GP (1994) The density and diversity of bird populations in three residential communities in Edmonton, Alberta. Can Field Nat 108:156–161
Emlen JT (1974) An urban bird community in Tucson, Arizona: derivation, structure, regulation. Condor 76:184–197
Escobar-Ibáñez JF, MacGregor-Fors I (2016) Peeking into the past to plan the future: assessing bird species richness in a Neotropical city. Urban Ecosyst 19:657–667
Evans KL, Newson SE, Gaston KJ (2009) Habitat influences on urban avian assemblages. Ibis 151:19–39
Faggi A, Perepelizin P (2006) Riqueza de aves a lo largo de un gradiente de urbanización en la ciudad de Buenos Aires. Rev Mus Argentino Cienc Nat 8:289–297
Faggi AM, Krellenberg K, Castro R et al (2006) Biodiversity in the Argentinean rolling Pampa Ecoregion: changes caused by agriculture and urbanisation. Erdkunde 60:127–138
Fauth JE, Bernardo J, Camara M et al (1996) Simplifying the jargon of community ecology: a conceptual approach. Am Nat 147:282–286
Fernández B, Zulma J, Antelo CM (2009) Composición y variaciones estacionales de la avifauna en un ambiente antropizado (San Pablo, Tucumán, Argentina). Acta Zool Lilloana 53:98–107
Fernández-Juricic E (2000) Avifaunal use of wooded streets in an urban landscape. Conserv Biol 14:513–521

Fernández-Juricic E (2001) Avian spatial segregation at edges and interiors of urban parks in Madrid, Spain. Biodivers Conserv 10:1303–1316
Filippi-Codaccioni O, Devictor V, Clobert J et al (2008) Effects of age and intensity of urbanization on farmland bird communities. Biol Conserv 141:2698–2707
Filippi-Codaccioni O, Clobert J, Julliard R (2009) Urbanisation effects on the functional diversity of avian agricultural communities. Acta Oecol 35:705–710
Fischer JD, Schneider SC, Ahlers AA et al (2015) Categorizing wildlife responses to urbanization and conservation implications of terminology. Conserv Biol 29:1246–1248
Fontana CS (2005) A ornitofauna de Porto Alegre no Século XX: Status de ocorrência e conservação. Comun Mus Ciênc Tecnol PUCRS Sér Zool 18:161–206
Fontana CS, Burger MI, Magnusson WE (2011) Bird diversity in a subtropical South-American City: effects of noise levels, arborisation and human population density. Urban Ecosyst 14:341–360
Fox S-JC, Hockey PAR (2007) Impacts of a South African coastal golf estate on shrubland bird communities. South Afr J Sci 103:27–34
Franz I, Cappelatti L, Barros MP (2010) Bird community in a forest patch isolated by the urban matrix at the Sinos River basin, Rio Grande do Sul State, Brazil, with comments on the possible local defaunation. Braz J Biol 70:1137–1148
Fraterrigo JM, Wiens JA (2005) Bird communities of the Colorado Rocky Mountains along a gradient of exurban development. Landsc Urban Plan 71:263–275
Fuller RA, Warren PH, Armsworth PR et al (2008) Garden bird feeding predicts the structure of urban avian assemblages. Divers Distrib 14:131–137
Fuller RA, Tratalos J, Gaston KJ (2009) How many birds are there in a city of half a million people? Divers Distrib 15:328–337
Gagné SA, Fahrig L (2011) Do birds and beetles show similar responses to urbanization? Ecol Appl 21:2297–2312
Garaffa PI, Filloy J, Bellocq MI (2009) Bird community responses along urban–rural gradients: does the size of the urbanized area matter? Landsc Urban Plan 90:33–41
Garitano-Zavala Á, Gismondi P (2003) Variación de la riqueza y diversidad de la ornitofauna en áreas verdes urbanas de las ciudades de La Paz y El Alto (Bolivia). Ecol Bolivia 38:65–78
Gaston KJ, Warren PH, Thompson K et al (2005) Urban domestic gardens (IV): the extent of the resource and its associated features. Biodivers Conserv 14:3327–3349
Gatesire T, Nsabimana D, Nyiramana A et al (2014) Bird diversity and distribution in relation to urban landscape types in Northern Rwanda. Sci World J 2014:157824
Gavareski CA (1976) Relation of park size and vegetation to urban bird populations in Seattle, Washington. Condor 78:375–382
Germain P, Cuevas Y, Sanhueza C et al (2008) Ensamble de aves en zonas con diferente grado de urbanización en la ciudad de Bahía Blanca (Buenos Aires, Argentina). Bioscriba 1:35–45
Germaine SS, Rosenstock SS, Schweinsburg RE et al (1998) Relationships among breeding birds, habitat, and residential development in Greater Tucson, Arizona. Ecol Appl 8:680–691
Glue D (2006) Variety at winter bird tables. Bird Popul 7:212–215
Gómez VE (2006) Aves del parque general San Martín (Mendoza). Distribución y características Multequina 15:81–95
González-García F, Straub R, Lobato García JA et al (2014) Birds of a neotropical green city: an up-to-date review of the avifauna of the city of Xalapa with additional unpublished records. Urban Ecosyst 17:991–1012
González-García F, Straub R, Lobato García JA et al (2016) Nuevos registros y notas adicionales comentadas sobre la avifauna de la ciudad de Xalapa, Veracruz, México. Acta Zool Mex 32:253–269
González-Oreja JA (2011) Birds of different biogeographic origins respond in contrasting ways to urbanization. Biol Conserv 144:234–242

González-Oreja JA, De La Fuente-Díaz-Ordaz AA, Hernández-Santín L et al (2012a) Can human disturbance promote nestedness? Songbirds and noise in urban parks as a case study. Landsc Urban Plan 104:9–18
González-Oreja JA, Barillas-Gómez AL, Bonache-Regidor C et al (2012b) Does habitat heterogeneity affect bird community structure in urban parks? Stud Avian Biol 45:1–16
González-Urrutia M (2009) Avifauna urbana en América Latina: Estudios de casos. Gest Ambient 17:55–68
Gotelli NJ, Colwell RK (2001) Quantifying biodiversity: procedures and pitfalls in the measurement and comparison of species richness. Ecol Lett 4:379–391
Hahs AK, McDonnell MJ (2006) Selecting independent measures to quantify Melbourne's urban–rural gradient. Landsc Urban Plan 78:435–448
Heil L, Fernández-Juricic E, Renison D et al (2007) Avian responses to tourism in the biogeographically isolated high Córdoba Mountains, Argentina. Biodivers Conserv 16:1009–1026
Holtz RE (1998) Avian species diversity in urban and suburban areas. Loon 70:40–42
Huang Y, Zhao Y, Li S et al (2015) The effects of habitat area, vegetation structure and insect richness on breeding bird populations in Beijing urban parks. Urban For Urban Green 14:1027–1039
Huhtalo H, Järvinen O (1977) Quantitative composition of the urban bird community in Tornio, northern Finland. Bird Study 24:179–185
Husté A, Selmi S, Boulinier T (2006) Bird communities in suburban patches near Paris: determinants of local richness in a highly fragmented landscape. Ecoscience 13:249–257
Iaquinta DL, Drescher AW (2000) Defining periurban: understanding rural-urban linkages and their connection to institutional contexts. Paper presented at the Tenth World Congress of the International Rural Sociology Association, Rio de Janeiro, 1 Aug 2000
Ikin K, Knight E, Lindenmayer DB et al (2013) The influence of native versus exotic streetscape vegetation on the spatial distribution of birds in suburbs and reserves. Divers Distrib 19:294–306
Ikin K, Barton PS, Knight E et al (2014) Bird community responses to the edge between suburbs and reserves. Oecologia 174:545–557
Jaccard P (1912) The distribution of the flora in the alpine zone. New Phytol 11:37–50
Johnston RF (2001) Synanthropic birds of north America. In: Marzluff JM, Bowman R, Donnelly R (eds) Avian ecology and conservation in an urbanizing world. Springer, New York, pp 49–67
Jokimäki J, Kaisanlahti-Jokimäki ML, Sorace A et al (2005) Evaluation of the "safe nesting zone" hypothesis across an urban gradient: a multi-scale study. Ecography 28:59–70
Jones DN (1981) Temporal changes in the suburban avifauna of an inland city. Wildl Res 8:109–119
Jones DN, Wieneke J (2000) The suburban bird community of Townsville revisited: changes over a 17 year period. Corella 24:53–60
Juri MD, Chani JM (2009) Variación estacional en la composición de las comunidades de aves en un gradiente urbano. Ecol Austral 19:175–184
Kang W, Minor ES, Park CR et al (2015) Effects of habitat structure, human disturbance, and habitat connectivity on urban forest bird communities. Urban Ecosyst 18:857–870
Kennedy CM, Marra PP, Fagan WF et al (2010) Landscape matrix and species traits mediate responses of Neotropical resident birds to forest fragmentation in Jamaica. Ecol Monogr 80:651–669
Klem D Jr (1989) Bird-window collisions. Wilson Bull 101:606–620
Knight RL, Gutzwiller KJ (eds) (1995) Wildlife and recreationists: coexistence through management and research. Island Press, Washington D.C
Koleff P, Gaston KJ, Lennon JJ (2003) Measuring beta diversity for presence–absence data. J Anim Ecol 72:367–382
Kopij G (2011) Avian diversity in rural and urbanized habitats in Lesotho. Berkut 20:22–28
Kopij G (2015) Avian diversity in an urbanized south African grassland. Zool Ecol 25:87–100
Korolkova GE (1978) Changes in the bird population of forest-park zones. In: Dunaeva TN, Isakov IA, Kazanskaya NS et al (eds) Plant and animal life in Moscow and its environs. Moscow University, Moscow, pp 55–56

Kristan WB, Lynam AJ, Price MV et al (2003) Alternative causes of edge-abundance relationships in birds and small mammals of California coastal sage scrub. Ecography 26:29–44
La Sorte FA, Tingley MW, Hurlbert AH (2014) The role of urban and agricultural areas during avian migration: an assessment of within-year temporal turnover. Glob Ecol Biogeogr 23:1225–1234
Lancaster RK, Rees WE (1979) Bird communities and the structure of urban habitats. Can J Zool 57:2358–2368
Le Roux DS, Ikin K, Lindenmayer DB et al (2015) Single large or several small? Applying biogeographic principles to tree-level conservation and biodiversity offsets. Biol Conserv 191:558–566
Lee M-B, Carroll JP (2014) Relative importance of local and landscape variables on site occupancy by avian species in a pine forest, urban, and agriculture matrix. For Ecol Manag 320:161–170
Lee D, Kim E, Choi J et al (2010) The effects of development on forest-patch characteristics and bird diversity in Suji, South Korea. Landsc Ecol Eng 6:171–179
Lepczyk CA, Flather CH, Radeloff VC et al (2008) Human impacts on regional avian diversity and abundance. Conserv Biol 22:405–416
Lerman SB, Turner VK, Bang C (2012) Homeowner associations as a vehicle for promoting native urban biodiversity. Ecol Soc 17:45
Leveau LM (2013) Relaciones aves–habitat en el sector suburbano de Mar del Plata, Argentina. Ornitol Neotrop 24:201–212
Leveau LM, Leveau CM (2004) Comunidades de aves en un gradiente urbano de la ciudad de Mar del Plata. El Hornero (B. Aires) 19:13–21
Leveau LM, Isla FI, Bellocq MI (2015) Urbanization and the temporal homogenization of bird communities: a case study in central Argentina. Urban Ecosyst 18:1461–1476
Lin S, Ye Y, Li J et al (2013) The study of bird diversity in Tiegang reservoir in the quickly urbanization area. Sichuan J Zool 32:297–301
Lopes EV, dos Anjos L (2006) A composição da avifauna do campus da Universidade Estadual de Londrina, norte do Paraná, Brasil. Rev Bras Zool 23:145–156
Loss SR, Ruiz MO, Brawn JD (2009) Relationships between avian diversity, neighborhood age, income, and environmental characteristics of an urban landscape. Biol Conserv 142:2578–2585
Luck GW, Smallbone LT, Sheffield KJ (2013) Environmental and socio-economic factors related to urban bird communities. Austral Ecol 38:111–120
Luniak M, Jabłoński P, Marczak P (1986) Birds of the Łazienki Krolewskie (Warsaw) in the years 1954-84. Acta Ornithol 22:23–97
Luther D, Hilty J, Weiss J et al (2008) Assessing the impact of local habitat variables and landscape context on riparian birds in agricultural, urbanized, and native landscapes. Biodivers Conserv 17:1923–1935
MacArthur RH, MacArthur JW (1961) On bird species diversity. Ecology 42:594–598
MacGregor-Fors I (2005) Listado ornitológico del Centro Universitario de Ciencias Biológicas y Agropecuarias, Universidad de Guadalajara, Jalisco, México: Un espacio suburbano. Huitzil 6:1–6
MacGregor-Fors I (2008) Relation between habitat attributes and bird richness in a western Mexico suburb. Landsc Urban Plan 84:92–98
MacGregor-Fors I (2010) How to measure the urban-wildland ecotone: redefining "peri-urban" areas. Ecol Res 25:883–887
MacGregor-Fors I, Ortega-Álvarez R (2011) Fading from the forest: bird community shifts related to urban park site-specific and landscape traits. Urban For Urban Green 10:239–246
MacGregor-Fors I, Ortega-Álvarez R (eds) (2013) Ecología Urbana: Experiencias en América Latina. Aviable via INECOL http://www1.inecol.edu.mx/libro_ecologia_urbana/. Accessed 18 Jan 2017
MacGregor-Fors I, Payton ME (2013) Contrasting diversity values: statistical inferences based on overlapping confidence intervals. PLoS One 8:e56794
MacGregor-Fors I, Schondube JE (2011) Gray vs. green urbanization: relative importance of urban features for urban bird communities. Basic Appl Ecol 12:372–381

MacGregor-Fors I, Morales-Pérez L, Quesada J et al (2010) Relationship between the presence of House Sparrows (*Passer domesticus*) and Neotropical bird community structure and diversity. Biol Invasions 12:87–96
MacGregor-Fors I, Morales-Pérez L, Schondube J (2012) From forests to cities: effects of urbanization on tropical birds. In: Lepczyk CA, Warren PS (eds) Urban bird ecology and conservation. University of California Press, California, pp 33–48
MacGregor-Fors I, Avendaño-Reyes S, Bandala VM et al (2015) Multi-taxonomic diversity patterns in a neotropical green city: a rapid biological assessment. Urban Ecosyst 18:633–647
Maeda T (1998) Bird communities and habitat relationships in a residential area of Tokyo. J Yamashina Inst Ornithol 30:83–100
Magurran AE (2005) Biological diversity. Curr Biol 15:R116–R118
Magurran AE, McGill BJ (2011) Biological diversity: Frontiers in measurement and assessment. Oxford University Press, New York
Malagamba-Rubio A, MacGregor-Fors I, Pineda-López R (2013) Comunidades de aves en áreas verdes de la ciudad de Santiago de Querétaro, México. Ornitol Neotrop 24:371–386
Manhães MA, Loures-Ribeiro A (2005) Spatial distribution and diversity of bird community in an urban area of Southeast Brazil. Braz Arch Biol Technol 48:285–294
Maragliano RE, Marti LJ, Ibañez LM et al (2009) Comunidades de aves urbanas de Lavallol, Buenos Aires, Argentina. Acta Zool Lilloana 53:108–114
Marín-Gómez OH (2005) Avifauna del campus de la Universidad del Quindío. Bol SAO XV:42–60
Marzluff JM, Bowman R, Donnelly R (2001) A historical perspective on urban bird research: trends, terms, and approaches. In: Marzluff JM, Bowman R, Donnelly R (eds) Avian ecology and conservation in an urbanizing world. Springer, New York, pp 1–17
Mason P (1985) The impact of urban development on bird communities of three Victorian towns-Lilydale, Coldstream and Mt Evelyn. Corella 9:14–21
McDonnell MJ, Pickett STA (1990) Ecosystem structure and function along urban-rural gradients: an unexploited opportunity for ecology. Ecology 71:1232–1237
McKinney ML (2006) Urbanization as a major cause of biotic homogenization. Biol Conserv 127:247–260
Melles S, Glenn S, Martin K (2003) Urban bird diversity and landscape complexity: species-environment associations along a multiscale habitat gradient. Conserv Ecol 7:5
Miller SG, Knight RL, Miller CK (1998) Influence of recreational trails on breeding bird communities. Ecol Appl 8:162–169
Mills GS, Dunning JB, Bates JM (1989) Effects of urbanization on breeding bird community structure in southwestern desert habitats. Condor 91:416–428
Molina D, Torres-Guerrero J, MDLL A-G (2012) Riqueza de aves del Área Natural Protegida estero El Salado, Puerto Vallarta, Jalisco, México. Huitzil 13:22–33
Monnerat-Nogueira D, Andrade-Rangel de Freitas A, Pinheiro da-Silva C et al (2005) Estudio de la avifauna y sus ectoparásitos en un fragmento de bosque Atlántico en la Ciudad del Río de Janeiro, Brasil. Bol SAO XV:26–36
Moreno CE, Rodríguez P (2010) A consistent terminology for quantifying species diversity? Oecologia 163:279–282
Moreno CE, Sánchez-Rojas G, Pineda E et al (2007) Shortcuts for biodiversity evaluation: a review of terminology and recommendations for the use of target groups, bioindicators and surrogates. Int J Environ Health 1:71–86
Morneau F, Décarie R, Pelletier R et al (1999) Changes in breeding bird richness and abundance in Montreal parks over a period of 15 years. Landsc Urban Plan 44:111–121
Mörtberg UM (2001) Resident bird species in urban forest remnants; landscape and habitat perspectives. Landsc Ecol 16:193–203
Muñoz MC, Fierro-Calderón K, Rivera-Gutierrez HF (2007) Las aves del campus de la Universidad del Valle, una isla verde urbana en Cali, Colombia. Ornitol Colomb 5:5–20
Murgui E (2007) Effects of seasonality on the species–area relationship: a case study with birds in urban parks. Glob Ecol Biogeogr 16:319–329

Murthy AC, Fristoe TS, Burger JR (2016) Homogenizing effects of cities on North American winter bird diversity. Ecosphere 7:e01216
Myers N, Mittermeier RA, Mittermeier CG et al (2000) Biodiversity hotspots for conservation priorities. Nature 403:853–858
Navarro C, Brandán Fernández Z, Marigliano N et al (2011) Birds associated to sections of riparian forests with anthropogenic modifications (Tucumán, Argentina): I. General characteristics. Acta Zool Lilloana 55:109–122
Nilon CH, Warren PS, Wolf J (2011) Baltimore birdscape study: identifying habitat and land-cover variables for an urban bird-monitoring project. Urban Habitats 6:1–9
Olden JD (2006) Biotic homogenization: a new research agenda for conservation biogeography. J Biogeogr 33:2027–2039
Ormond SE, Whatmough R, Hudson I et al (2014) Environmental and anthropogenic impacts on avifaunal assemblages in an urban parkland, 1976 to 2007. Animals 4:119–130
Ortega-Álvarez R, MacGregor-Fors I (2009) Living in the big city: effects of urban land-use on bird community structure, diversity, and composition. Landsc Urban Plan 90:189–195
Ortega-Álvarez R, MacGregor-Fors I (2011a) Spreading the word: the ecology of urban birds outside the United States, Canada, and Western Europe. Auk 128:415–418
Ortega-Álvarez R, MacGregor-Fors I (2011b) Dusting-off the file: a review of knowledge on urban ornithology in Latin America. Landsc Urban Plan 101:1–10
Paker Y, Yom-Tov Y, Alon-Mozes T et al (2014) The effect of plant richness and urban garden structure on bird species richness, diversity and community structure. Landsc Urban Plan 122:186–195
Palita SK, Ponkshe AV, Dhar U (2011) Habitat enrichment and its impact on avian diversity: a study at GBPIHED, Kosi-Katarmal, Uttarakhand, India. Curr Sci 100:1681–1689
Palmer GC, Fitzsimons JA, Antos MJ et al (2008) Determinants of native avian richness in suburban remnant vegetation: implications for conservation planning. Biol Conserv 141:2329–2341
Palomino D, Carrascal LM (2005) Birds on novel island environments. A case study with the urban avifauna of Tenerife (Canary Islands). Ecol Res 20:611–617
Parsons H, French K, Major R (2003) The influence of remnant bushland on the composition of suburban bird assemblages in Australia. Landsc Urban Plan 66:43–56
Pautasso M, Dinetti M (2009) Avian species richness, human population and protected areas across Italy's regions. Environ Conserv 36:22–31
Pennington DN, Hansel J, Blair RB (2008) The conservation value of urban riparian areas for landbirds during spring migration: land cover, scale, and vegetation effects. Biol Conserv 141:1235–1248
Platt A, Lill A (2006) Composition and conservation value of bird assemblages of urban "habitat islands": do pedestrian traffic and landscape variables exert an influence? Urban Ecosyst 9:83–97
Poague KL, Johnson RJ, Young LJ (2000) Bird use of rural and urban converted railroad rights-of-way in southeast Nebraska. Wildl Soc Bull 28:852–864
Presti PM, Echevarria AL (2009) El ensamble de aves en un relicto de selva pedemontana: Parque Percy Hill (Yerba Buena, Tucumán, Argentina). Acta Zool Lilloana 53:64–76
Puga-Caballero A, MacGregor-Fors I, Ortega-Álvarez R (2014) Birds at the urban fringe: Avian community shifts in different peri-urban ecotones of a megacity. Ecol Res 29:619–628
Reynaud PA, Thioulouse J (2000) Identification of birds as biological markers along a neotropical urban-rural gradient (Cayenne, French Guiana), using co-inertia analysis. J Environ Manag 59:121–140
Rivera-Gutiérrez HF (2006) Composición y estructura de una comunidad de aves en un área suburbana en el suroccidente colombiano. Ornitol Colomb 4:28–38
Rodríguez-Estrella R (2007) Land use changes affect distributional patterns of desert birds in the Baja California peninsula, Mexico. Divers Distrib 13:877–889
Rolando A, Maffei G, Pulcher C et al (1997) Avian community structure along an urbanization gradient. Ital J Zool 64:341–349

Rosa S, Palmeirim JM, Moreira F (2003) Factors affecting waterbird abundance and species richness in an increasingly urbanized area of the Tagus estuary in Portugal. Waterbirds 26:226–232
Rosselli L, Stiles FG (2012) Local and landscape environmental factors are important for the conservation of endangered wetland birds in a high Andean plateau. Waterbirds 35:453–469
Sacchi R, Gentilli A, Razzetti E et al (2002) Effects of building features on density and flock distribution of feral pigeons *Columba livia* var. *domestica* in an urban environment. Can J Zool 80:48–54
Sacco AG, Rui AM, Bergmann FB et al (2015) Perda de diversidade taxonômica e funcional de aves em área urbana no sul do Brasil. Iheringia Ser Zool 105:276–287
Santos KC, Pino J, Rodà F et al (2008) Beyond the reserves: the role of non-protected rural areas for avifauna conservation in the area of Barcelona (NE of Spain). Landsc Urban Plan 84:140–151
Sanz V, Caula S (2015) Assessing bird assemblages along an urban gradient in a Caribbean island (Margarita, Venezuela). Urban Ecosyst 18:729–746
Scherer A, Scherer SB, Bugoni L et al (2005) Estrutura trófica da avifauna em oito parques da cidade de Porto Alegre, Rio Grande do Sul, Brasil. Ornithologia 1:25–32
Schneider SC, Miller JR (2014) Response of avian communities to invasive vegetation in urban forest fragments. Condor 116:459–471
Schneider SC, Fischer JD, Miller JR (2015) Two-sided edge responses of avian communities in an urban landscape. Urban Ecosyst 18:539–551
Schwarzenberger A, Dean W (2003) The influence of vegetation structure on bird communities in a Karoo village, South Africa. Ostrich J Afr Ornithol 74:209–216
Sengupta S, Mondal M, Basu P (2014) Bird species assemblages across a rural urban gradient around Kolkata, India. Urban Ecosyst 17:585–596
Shanahan DF, Miller C, Possingham HP et al (2011) The influence of patch area and connectivity on avian communities in urban revegetation. Biol Conserv 144:722–729
Shultz AJ, Tingley MW, Bowie RCK (2012) A century of avian community turnover in an urban green space in Northern California. Condor 114:258–267
Shwartz A, Shirley S, Kark S (2008) How do habitat variability and management regime shape the spatial heterogeneity of birds within a large Mediterranean urban park? Landsc Urban Plan 84:219–229
Silva FD, Blamires D (2007) Avifauna urbana no Lago Pôr do Sol, Iporá, Goiás, Brasil. Lundiana Int J Biodivers 8:17–26
Silva CP, García CE, Estay SA et al (2015) Bird richness and abundance in response to urban form in a Latin American city: Valdivia, Chile as a case study. PLoS One 10:e0138120
Smith PG (2003) Winter bird use of urban and rural habitats in Ontario. Can Field Nat 117:173–183
Smith LA, Chow-Fraser P (2010) Impacts of adjacent land use and isolation on marsh bird communities. Environ Manag 45:1040–1051
Smith CM, Wachob DG (2006) Trends associated with residential development in riparian breeding bird habitat along the Snake River in Jackson Hole, WY, USA: implications for conservation planning. Biol Conserv 128:431–446
Snep RPH, Opdam PFM, Baveco JM et al (2006) How peri-urban areas can strengthen animal populations within cities: a modeling approach. Biol Conserv 127:345–355
Soares AG (2004) A avifauna de uma Area no bairro Ribeirão da Ilha, Florianópolis, Santa Catarina: Levantamento e implicações para a educação ambiental. Biotemas 17:107–124
Sodhi NS (1992) Comparison between urban and rural bird communities in prairie Saskatchewan: urbanization and short-term population trends. Can Field Nat 106:210–215
Soh MCK, Sodhi NS, Lim SLH (2006) High sensitivity of montane bird communities to habitat disturbance in Peninsular Malaysia. Biol Conserv 129:149–166
Sol D, González-Lagos C, Moreira D et al (2014) Urbanisation tolerance and the loss of avian diversity. Ecol Lett 17:942–950
Song W (2015) Analysis of bird species diversity response to structural conditions of urban park. J Korean Env Rest Tech 18:65–77

Sorace A (2001) Value to wildlife of urban-agricultural parks: a case study from Rome urban area. Environ Manag 28:547–560
Strohbach MW, Hrycyna A, Warren PS (2014) 150 years of changes in bird life in Cambridge, Massachusetts from 1860 to 2012. Wilson J Ornithol 126:192–206
Suarez-Rubio M, Thomlinson JR (2009) Landscape and patch-level factors influence bird communities in an urbanized tropical island. Biol Conserv 142:1311–1321
Suk HY, Chung O-S, Lee J-Y et al (2014) Dynamic influence of patch size on occupancy of woodland birds. Anim Cells Syst 18:68–75
Sun Y, Lu CL, Wang Z et al (2015) Avian community structure and seasonal dynamic in Changguangxi national urban wetland park. Sichuan J Zool 34:541–547
Sushinsky JR, Rhodes JR, Possingham HP et al (2013) How should we grow cities to minimize their biodiversity impacts? Glob Change Biol 19:401–410
Takeuchi M, Kojima H, Watanabe M (2010) Avifauna at the Setagaya campus of the Tokyo University of Agriculture. J Agric Sci Tokyo Nogyo Daigaku 55:115–122
Tampson VE, Petry MV (2008) Nidificação e análise das guildas alimentares de aves no Morro do Espelho, na zona urbana de São Leopoldo–RS. Biodivers Pampeana 6:63–69
Teles AS, Blamires D, Dos Santos Reis ES (2012) Comunidade de aves no clube recreativo da SANEAGO em Iporá, Estado de Goiás. Biotemas 25:195–204
Tjallingii SP (2000) Ecology on the edge: landscape and ecology between town and country. Landsc Urban Plan 48:103–119
Toledo MCB, Donatelli RJ, Batista GT (2012) Relation between green spaces and bird community structure in an urban area in Southeast Brazil. Urban Ecosyst 15:111–131
Tomiałojć L (1998) Breeding bird densities in some urban versus non-urban habitats: the Dijon case. Acta Ornithol 33:159–171
Trammell EJ, Weisberg PJ, Bassett S (2011) Avian response to urbanization in the arid riparian context of Reno, USA. Landsc Urban Plan 102:93–101
Tryjanowski P, Sparks TH, Biaduń W et al (2015) Winter bird assemblages in rural and urban environments: a national survey. PLoS One 10:e0130299
Turner WR (2003) Citywide biological monitoring as a tool for ecology and conservation in urban landscapes: the case of the Tucson Bird Count. Landsc Urban Plan 65:149–166
United Nations (2015) World urbanization prospects – 2014 revision. United Nations, New York
Vallejo BM, Aloy AB, Ong PS (2009) The distribution, abundance and diversity of birds in Manila's last greenspaces. Landsc Urban Plan 89:75–85
van Heezik Y, Smyth A, Mathieu R (2008) Diversity of native and exotic birds across an urban gradient in a New Zealand city. Landsc Urban Plan 87:223–232
van Rensburg BJ, Peacock DS, Robertson MP (2009) Biotic homogenization and alien bird species along an urban gradient in South Africa. Landsc Urban Plan 92:233–241
Verma SK, Murmu TD (2015) Impact of environmental and disturbance variables on avian community structure along a gradient of urbanization in Jamshedpur, India. PLoS One 10:e0133383
Vignoli L, Scirè S, Bologna MA (2013) Rural–urban gradient and land use in a millenary metropolis: how urbanization affects avian functional groups and the role of old villas in bird assemblage patterning. Web Ecol 13:49–67
Villegas M, Garitano-Zavala Á (2010) Bird community responses to different urban conditions in La Paz, Bolivia. Urban Ecosyst 13:375–391
Vogel HF, Metri R, Zawadzki CH et al (2011) Avifauna from a campus of Universidade Estadual do Centro-Oeste, Guarapuava, Paraná State, Brazil. Acta Sci Biol Sci 33:197–207
Walker JS, Shochat E (2010) Scalar effects of vegetation on bird communities in an urbanizing desert ecosystem. Urban Ecosyst 13:155–167
White JG, Antos MJ, Fitzsimons JA et al (2005) Non-uniform bird assemblages in urban environments: the influence of streetscape vegetation. Landsc Urban Plan 71:123–135
Wood EM, Pidgeon AM, Radeloff VC et al (2015) Long-term avian community response to housing development at the boundary of US protected areas: effect size increases with time. J Appl Ecol 52:1227–1236

Yang G, Xu J, Wang Y et al (2015a) Evaluation of microhabitats for wild birds in a Shanghai urban area park. Urban For Urban Green 14:246–254

Yang G, Xu J, Wang Y et al (2015b) The influence of vegetation structure on bird guilds in an urban park. Acta Ecol Sin 35:4824–4835

Yeoman F, Nally RM (2005) The avifaunas of some fragmented, periurban, coastal woodlands in south-eastern Australia. Landsc Urban Plan 72:297–312

Young KM, Daniels CB, Johnston G (2007) Species of street tree is important for southern hemisphere bird trophic guilds. Austral Ecol 32:541–550

Zhang L, Miao Y (2013) Avian community diversity in Urumqi downtown parks and green lands. Sichuan J Zool 32:97–102

Zhou D, Chu LM (2014) Do avian communities vary with season in highly urbanized Hong Kong? Wilson J Ornithol 126:69–80

Chapter 4
Flocking the City: Avian Demography and Population Dynamics in Urban Latin America

Lucas M. Leveau and Iriana Zuria

Abstract We here provide a general overview of the study of bird abundance, demography, and populations in urban environments with special attention on those from Latin America. We found that studies focused on bird abundance and demography were more common in Argentina, Brazil, and Mexico. We classified those species taken into account in four or more studies according to their response to urbanization. In general, specialist birds are negatively related with urbanization, whereas two invasive exotics, the Rock Pigeon (*Columba livia*) and House Sparrow (*Passer domesticus*), are present in most of the reviewed urbanized areas along Latin America. Although scarce, studies of bird demography revealed, in general, higher bird survival in urban environments, possibly due to lower predator density. However, anthropogenic causes have been identified to affect bird populations in urban areas (e.g., window and vehicle collisions, nest predation, predation by domestic cats and dogs). Future studies will importantly add to our knowledge of urban bird ecology in Latin America by continuing to quantify bird numbers and assess demography patterns, as well as identifying the factors that mold them. Birds, as highly informative and charismatic bioindicators, could play a crucial role in adding evidence-based knowledge for decision-makers to take action in building bird-friendly and livable cities.

Keywords Avian populations • Breeding success • Density • Metapopulations • Survival

L.M. Leveau
Departamento de Ecología, Genética y Evolución, Universidad de Buenos Aires – IEGEBA (CONICET-UBA), Intendente Güiraldes 2160, Ciudad Universitaria, Pab 2, Piso 4, C1428EGA Buenos Aires, Argentina

I. Zuria (✉)
Centro de Investigaciones Biológicas, Universidad Autónoma del Estado de Hidalgo, Km 4.5 carr. Pachuca-Tulancingo s/n Ciudad del Conocimiento, Mineral de la Reforma, Hidalgo, CP 42184, Mexico
e-mail: izuria@uaeh.edu.mx

I. MacGregor-Fors, J.F. Escobar-Ibáñez (eds.), *Avian Ecology in Latin American Cityscapes*, DOI 10.1007/978-3-319-63475-3_4

4.1 Introduction

Understanding the limitations for bird populations is fundamental, as such information provides the foundation for practical wildlife management actions (Newton 1998). Urban areas are markedly different from nonurban ones in many ways, including resource flows, pollution, and species interactions (Shanahan et al. 2014). As a consequence of such drastic differences, urbanization constitutes a compound filter for bird species, with some species unable to dwell in urban environments and others that thrive within them (Blair 1996; Croci et al. 2008; MacGregor-Fors et al. 2010a; Fischer et al. 2015).

The aim of this chapter is to explore our current knowledge of urban bird abundance, demography, and population dynamics with an important focus on Latin America. For this, we performed an intensive search of urban bird studies in the Web of Science (www.webofknowldge.com) and Scopus (www.scopus.com) platforms using keyword combinations including the terms 'urban' and 'bird', as well as the names of all Latin American countries (see Chaps. 2 and 3 for a list of countries). We then focused on studies that reported information on bird abundances (in any of its expressions; e.g., density, number of individuals) or demographic parameters within urban settings. Based on the retrieved information, we set a global perspective regarding avian abundance and demography patterns in urban systems. Finally, we narrowed our analysis to Latin America to finally discuss current research gaps and potential future directions in this particular topic in the region.

4.2 Global Perspective

4.2.1 Abundance

One of the most consistent urban bird ecology patterns across the world is the increase of total bird numbers with urbanization, contrasting with important species richness declines (Marzluff et al. 2001; Chace and Walsh 2006; Ortega-Álvarez and MacGregor-Fors 2011; Aronson et al. 2014). Although avian abundance can be a misleading indicator of habitat quality (Van Horne 1983; Bock and Jones 2004), it often constitutes the primary indicator of species responses to urbanization, having important conservation and management implications (Marzluff et al. 2001; Rodewald and Gehrt 2014). Species numbers have been commonly used in urban studies as a baseline to classify species in different categories according to their response to urbanization intensity (Blair 1996; McKinney 2002; Rodewald and Gehrt 2014; Fischer et al. 2015).

Certainly, a number of environmental factors such as the availability of resources, as well as the presence of competing species and natural enemies, both predators and parasites (Newton 1998), can determine the differential responses of bird species to urbanization. For example, food supplementation is a typical urban-related alteration of resource availability that can enhance bird productivity (Robb et al. 2008; Amrhein 2014), although bird aggregation around feeders may attract predators and/or promote the spread of diseases (Robb et al. 2008; Martin and Boruta 2014). In fact, higher bird densities in urban areas have been associated with lower predation, suggesting that urbanization reduces top-down control, but may trigger bottom-up processes through the greater availability of anthropogenic food resources (Shochat et al. 2006; Fischer et al. 2012). Also, urbanization can alter interspecific interactions by benefiting aggressive species, such as the House Sparrow (*Passer domesticus*) in North America (Gowaty 1984; MacGregor-Fors et al. 2010b) and Noisy Miner (*Manorina melanocephala*) in suburban areas of Australia (Catterall 2004; Parsons et al. 2006; Montague-Drake et al. 2011), which often displace other native species from urban systems.

Species that reach their highest numbers in highly urbanized areas (sensu MacGregor-Fors 2011) are commonly named urban 'exploiters', urban 'dependents', or urban 'dwellers' (Blair 1996; McKinney 2002; Rodewald and Gehrt 2014; Fischer et al. 2015). Classic examples of these species around the world are the Rock Pigeon (*Columba livia*), the House Sparrow, and the European Starling (*Sturnus vulgaris*), all highly adapted to urban environments (Clergeau et al. 2006; Kark et al. 2007; Blair and Johnson 2008; Aronson et al. 2014). A second group of species that can thrive in urban areas, but often depend on nonurban populations, are the so-called urban 'adapters' or 'users' (Blair 1996; McKinney 2002; Fischer et al. 2015). As the term 'adapter' has been criticized because it has an erroneous evolutionary connotation and potential mismatches between density and habitat quality (Rodewald and Gehrt 2014), we here refer to them as urban 'users' (sensu Fischer et al. 2015). Some biological and behavioral traits of these species include granivorous and omnivorous diets, ground-foraging habits, and tree and shrub nesting (Beissinger and Osborne 1982; Merenlender et al. 2009; Ikin et al. 2014). Iconic urban users from North America include the American Robin (*Turdus migratorius*) and the House Finch (*Haemorhous mexicanus*), whereas their European analogs are the Common Blackbird (*Turdus merula*) and the Common Chaffinch (*Fringilla coelebs*) (Blair 2004; Caula et al. 2008). Finally, a third group of species, with highest numbers in nonurban areas and only present in low-intensity urbanized environments, has been called urban 'avoiders' (Blair 1996; Rodewald and Gehrt 2014; Fischer et al. 2015). Typical biological and behavioral traits of these species include an insectivorous diet, ground-nesting and solitary habits, and migratory behavior (Kark et al. 2007; Blair and Johnson 2008; Croci et al. 2008; Conole and Kirkpatrick 2011; Evans et al. 2011).

4.2.2 Demography

In 2001, Marzluff et al. published a pioneer and historical perspective on urban bird research. At that time, the majority of studies were based on correlative associations between the relative abundance of several breeding bird species and urbanization intensity. Since then, many researchers from across the globe have studied urban bird demography, including topics as survival (e.g., Mannan et al. 2004; Whittaker and Marzluff 2009; Ausprey and Rodewald 2011; Evans et al. 2015), breeding success (e.g., Phillips et al. 2005; Rodewald and Shustack 2008a), and mortality (e.g., Baker et al. 2008; Hager et al. 2008; Loss et al. 2014a, b).

Many of these studies, focused on the demographic parameters of urban birds, have shown contradictory results. For example, some studies have suggested weak relationships between adult survival rates and urbanization (Rodewald and Shustack 2008a, b; Evans et al. 2015). For some species, adult survival can even increase with urbanization, as happens with the Great Tit (*Parus major*) in Estonia (Horak and Lebreton 1998). Interestingly, in a study from Washington, DC, some species (i.e., Northern Cardinal—*Cardinalis cardinalis*, Gray Catbird—*Dumetella carolinensis*, American Robin, Song Sparrow—*Melospiza melodia*) showed peak survival rates at intermediate and highest levels of urbanization (Evans et al. 2015).

Regarding the role of urbanization on the breeding success of landbirds, studies have also shown differing results, with studies reporting higher, lower, and null reproductive productivity associated with urbanization (e.g., Phillips et al. 2005; Conway et al. 2006; Leston and Rodewald 2006; Rodewald and Shustack 2008a; Sumasgutner et al. 2014; Weaving et al. 2016). Contrasting and inconclusive results emphasize the need for more studies that help to understand the mechanisms that drive avian demography in urbanized landscapes (Evans et al. 2015). Besides the mortality associated to the nesting process, many studies on bird mortality in urban areas have focused on predation by domestic cats, bird–vehicle collisions, and bird–window collisions (see Chap. 7 for a thorough review and analysis on urban threats to birds). In summary, the two main causes of avian mortality have been shown to be collisions with man-made structures and predation by cats. Avian collision, especially with building windows, is estimated to be the cause of death of up to one billion birds annually in the United States alone (Loss et al. 2014b). Although some bird species may be experiencing population declines due to this cause (Klem 1990; Loss et al. 2015), a recent meta-analysis has suggested no causal relationships between bird collisions and population declines (Arnold and Zink 2011). Regarding the role of cats on bird populations, studies have suggested that direct predation is a major driver of avian populations (Baker et al. 2008; van Heezik et al. 2010; Loss et al. 2012), as well as some indirect effects, such as cat presence at bird nest sites reducing subsequent parental provisioning rates to chicks (Bonnington et al. 2013), which can in turn decrease nestling growth rates and decrease reproductive success (Dunn et al. 2010; Martin et al. 2011).

4.3 Latin America

4.3.1 Abundance

We gathered a total of 42 studies that report quantitative information that allow for the comparison of bird species abundance in sites with different urbanization intensities of Latin America. A large proportion of the studies were performed in Argentina (36%), Brazil (17%), and Mexico (17%), which in general echoes with the pattern reported in a previous general review focused on urban ornithology in Latin America (Ortega-Álvarez and MacGregor-Fors 2011). Most studies (71%) comparing bird numbers among sites with different urbanization intensities were located within the urban matrix (Varela-Contador 2003; Leveau and Leveau 2004; MacGregor-Fors et al. 2010a; Silva et al. 2016); yet, others were conducted in urban greenspaces (Toledo 2007; Suarez-Rubio and Thomlinson 2009; Carbó-Ramírez and Zuria 2011; González-Oreja et al. 2012). Fewer studies compared bird abundances among urbanization intensity scenarios, for instance: (i) between urban greenspaces and nonurban areas (Carvalho and Marini 2007; Kennedy et al. 2010; Chávez-Zichinelli et al. 2013); (ii) at increasing distances from urban settlements (Thornton et al. 2012; Lozano and Malo 2013); (iii) and in urban wetlands (Villaneda-Rey and Rosselli 2011; Rosselli and Stiles 2012). Although most of the studies were performed in a single urban area, a couple of studies were conducted in three or more cities (Thornton et al. 2012; Campbell 2014). Given that the processes that may determine bird abundances within the urban matrix can operate differently from those affecting birds found only in urban greenspaces (Rodewald and Gehrt 2014), we here deepen in the analysis of the response of birds to urbanization within urban matrices and greenspaces.

4.3.1.1 Urban Matrix

According to their abundance shifts related to urbanization intensity, we classified bird species in three groups: (i) species with greater relative abundances in the nonurban systems or showing negative responses to urbanization; (ii) species with greater numbers in intermediate levels of urbanization; and (iii) species with largest abundances at most urbanized sites (see Fig. 4.1 for an example of three bird species from Mar del Plata, Argentina). In this chapter, we followed a modified version of the classification proposed by Marzluff et al. (2001) to align the different studied designs, which allowed us to discriminate among several kinds of urbanization intensity scenarios. Qualitatively, we refer to: (i) 'urban' areas as those dominated by buildings and with the highest vehicle and pedestrian traffic; (ii) 'suburban' areas as those mainly composed by houses with gardens; (iii) 'peri-urban' areas as those comprised mostly of houses with gardens located at the fringe of the city; (iv) 'exurban' areas as open land with scattered houses located outside the city fringe; (v) 'nonurban' as those with scarce to no houses and with human activities such as

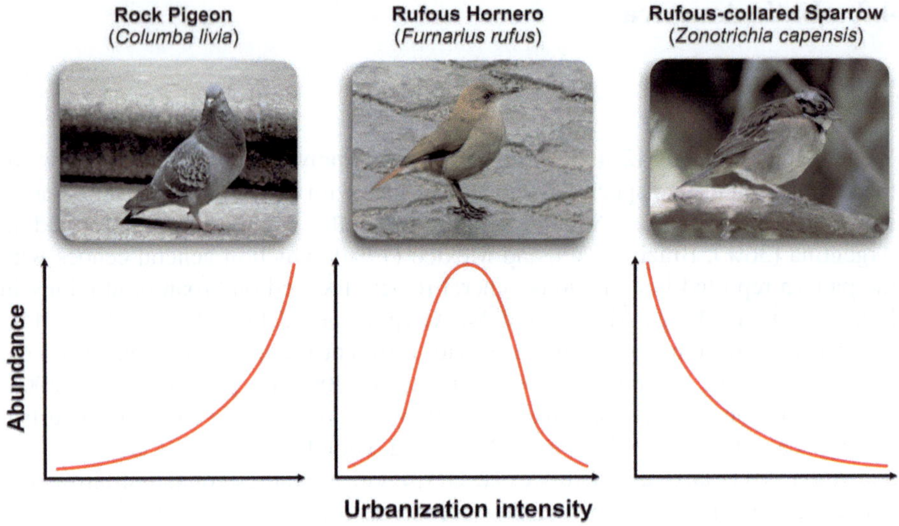

Fig. 4.1 Bird abundance shifts examples along an urbanization intensity gradient of Mar del Plata, Argentina (Patterns from: Leveau and Leveau 2004, 2005) (Photos: Ian MacGregor-Fors–Rock Pigeon; LML–Rufous Hornero; Juan F. Escobar-Ibáñez–Rufous-collared Sparrow)

agriculture or livestock grazing; and (vi) 'natural' areas as those where human presence and disturbance are low. We analyzed the type of response to urbanization, positive, negative, or null, for the 17 species recorded in four or more studies.

Most of the assessed species ($n = 9$) showed a positive response to scenarios with intermediate levels of urbanization (Table 4.1). Some clear examples of this pattern are the Eared Dove (*Zenaida auriculata*), Great Kiskadee (*Pitangus sulphuratus*), and House Wren (*Troglodytes aedon*). These species showed highest densities in 'suburban' or 'peri-urban' areas of several cities of Argentina, Brazil, Chile, Mexico, and Venezuela (Table 4.1). Most of these species are tree nesters and feed on the ground (Leveau and Leveau 2005; Leveau 2013a). Other species such as the Grayish Baywing (*Agelaioides badius*) and Tropical Kingbird (*Tyrannus melancholicus*) showed mixed responses to urbanization, both using and avoiding 'suburban' areas in different cities (Table 4.1). The Chimango Caracara (*Milvago chimango*) seems to be the only raptor species that consistently thrives in urban areas across its distribution in austral South America. The two exotic species included in our analysis (i.e., Rock Pigeon and House Sparrow) showed positive responses to increasing building cover at several spatial scales (Table 4.2), which are consistent with previous studies outside Latin America (Germaine et al. 1998; Blair 2004). Seven species had greater abundances in 'nonurban' and 'natural' areas when compared to developed ones, or were negatively related with urbanization (Table 4.3). Many of these species, such as the Southern Lapwing (*Vanellus chilensis*), Rufous-collared Sparrow (*Zonotrichia capensis*), and Grassland Yellow-Finch (*Sicalis luteola*), are typical grassland ground-nesters, making them particularly vulnerable to the land-use changes associated with urbanization (Leveau and Leveau 2005; Leveau 2013a).

Table 4.1 Relationship between bird numbers and urbanization for particular species responding generally positively to intermediate urbanization intensities in Latin American cities

Species	Study location	Response	Contrast[a]	Source[b]	Biome[c]
Picazuro Pigeon *Patagioenas picazuro*	Santa Fé Province, Argentina	–	S–N	1	FGS
	Mar del Plata, Argentina	–	S	2	TGSS
	Mar del Plata, Argentina	+	S–NU	3	TGSS
	Buenos Aires Province, Argentina	+	S	4	TGSS
Eared Dove *Zenaida auriculata*	Santa Fé Province, Argentina	+	S–N	1	FGS
	Tafi Viejo, Argentina	+	S–E	5	TSMBF
	San Miguel de Tucumán, Argentina	–	U–P	6	TSMBF
	Mar del Plata, Argentina	–	S	2	TGSS
	Mar del Plata, Argentina	+	S–NU	3	TGSS
	Buenos Aires Province, Argentina	+	S	4	TGSS
	Santiago de Chile, Chile	Null	S	7	MFWS
Chimango Caracara *Milvago chimango*	Junin, Argentina	–	U–NU	8	TGSS
	Mar del Plata, Argentina	–	S	2	TGSS
	Mar del Plata, Argentina	+	S–NU	3	TGSS
	Bariloche, Argentina	+	P–NU	9	TGSS
	Santiago de Chile, Chile	–	S	7	MFWS
	Temuco, Chile	+	U–NU	10	MFWS
	Valdivia, Chile	+	U–NU	10	TBMF
	Osorno, Chile	Null	U–NU	10	TBMF
Rufous Hornero *Furnarius rufus*	Santa Fé Province, Argentina	+	S–N	1	FGS
	Mar del Plata, Argentina	–	S	2	TGSS
	San Miguel de Tucumán, Argentina	–	U–P	6	TSMBF
	Mar del Plata, Argentina	+	S–NU	3	TGSS
	Buenos Aires Province, Argentina	–	S	8	TGSS
Great Kiskadee *Pitangus sulphuratus*	Santa Fé Province, Argentina	+	S–N	1	FGS
	Tafi Viejo, Argentina	+	S–E	5	TSMBF
	Mar del Plata, Argentina	–	S	2	TGSS
	Mar del Plata, Argentina	+	S–NU	3	TGSS
	San Miguel de Tucumán, Argentina	–	U–P	6	TSMBF
	Uberlândia, Brasil	Null	U–P	11	TSGSS
	Cayenne, French Guiana	Null	U–N	12	MFWS

(continued)

Table 4.1 (continued)

Species	Study location	Response	Contrast[a]	Source[b]	Biome[c]
House Wren *Troglodytes aedon*	Santa Fé Province, Argentina	+	S–N	1	FGS
	Tafi Viejo, Argentina	+	S–E	5	TSMBF
	San Miguel de Tucumán, Argentina	–	U–P	6	TSMBF
	Mar del Plata, Argentina	–	S	2	TGSS
	Mar del Plata, Argentina	+	S–NU	3	TGSS
	Buenos Aires Province, Argentina	Null	S	4	TGSS
	Santiago de Chile, Chile	Null	S	7	MFWS
	Temuco, Chile	–	U–NU	10	MFWS
	Valdivia, Chile	–	U–NU	10	TBMF
	Osorno, Chile	–	U–NU	10	TBMF
	Cayenne, French Guiana	Null	U–N	12	M
Rufous-bellied Thrush *Turdus rufiventris*	Santa Fé Province, Argentina	–	S–N	1	FGS
	Mar del Plata, Argentina	–	S	2	TGSS
	Mar del Plata, Argentina	+	S–NU	3	TGSS
	Tafi Viejo, Argentina	+	S–E	5	TSMBF
Grayish Baywing *Agelaioides badius*	Santa Fé Province, Argentina	+	S–N	1	FGS
	San Miguel de Tucumán, Argentina	–	U–P	6	TSMBF
	Mar del Plata, Argentina	–	S	2	TGSS
	Mar del Plata, Argentina	–	S–NU	3	TGSS
	Buenos Aires Province, Argentina	Null	S	4	TGSS
Shiny Cowbird *Molothrus bonariensis*	Santa Fé Province, Argentina	+	S–N	1	FGS
	San Miguel de Tucumán, Argentina	–	U–P	6	TSMBF
	Mar del Plata, Argentina	Null	S	2	TGSS
	Mar del Plata, Argentina	Null	S–NU	3	TGSS
	Buenos Aires Province, Argentina	+	S	4	TGSS
	Santiago de Chile, Chile	Null	S	7	MFWS
	Cayenne, French Guiana	–	U–N	12	M
	Ceiba, Puerto Rico	+	U–N	13	M

[a]*P–NU* Peri-urban–Nonurban, *S* Suburban, *S–E* Suburban–Ex-urban, *S–N* Suburban–Natural, *S–NU* Suburban–Nonurban, *U–N* Urban–Natural, *U–P* Urban–Peri-urban, *U–NU* Urban–Nonurban

[b]1: Rossetti and Giraudo (2003); 2: Leveau (2013b); 3: Leveau and Leveau (2005); 4: Leveau and Leveau (2006); 5: Navarro and Antelo (2014); 6: Juri and Chani (2005); 7: Varela-Contador (2003); 8: Bellocq et al. (2008); 9: Barbar et al. (2015); 10: Silva et al. (2016); 11: Amâncio et al. (2008); 12: Reynaud and Thioioulouse (2000); 13: Komar et al. (2012)

[c]*FGS* flooded grasslands and savannas, *M* mangroves, *MFWS* mediterranean forests, woodlands and scrub, *TBMF* temperate broadleaf and mixed forests, *TGSS* temperate grasslands, savannas, and shrublands, *TSGSS* tropical and subtropical grasslands, savannas, and shrublands, *TSMBF* tropical and subtropical moist broadleaf forests. Classification based on Olson et al. (2001)

Table 4.2 Relationship between bird abundances and urbanization for particular species responding generally positively to highly urbanized areas of Latin American cities

Species	Study location	Response	Contrast[a]	Source[b]	Biome[c]
Rock Pigeon *Columba livia*	Mar del Plata, Argentina	+	S	1	TGSS
	Mar del Plata, Argentina	+	S–NU	2	TGSS
	Buenos Aires Province, Argentina	+	S	3	TGSS
	San Miguel de Tucumán, Argentina	+	U–P	4	TSMBF
	Santiago de Chile, Chile	Null	S	5	MFWS
	Temuco, Chile	+	U–NU	6	MFWS
	Valdivia, Chile	+	U–NU	6	TBMF
	Osorno, Chile	+	U–NU	6	TBMF
	Uberlândia, Brasil	+	U–P	7	TSGSS
	Ceiba, Puerto Rico	+	U–N	8	M
	Careyes, Mexico	+	S–N	9	TSDBF
House Sparrow *Passer domesticus*	Santa Fé Province, Argentina	+	S–N	10	FGS
	San Miguel de Tucumán, Argentina	+	U–P	4	TSMBF
	Mar del Plata, Argentina	+	S	1	TGSS
	Mar del Plata, Argentina	+	S–NU	2	TGSS
	Santiago de Chile, Chile	+	S	5	MFWS
	Temuco, Chile	+	U–NU	6	MFWS
	Valdivia, Chile	+	U–NU	6	TBMF
	Osorno, Chile	+	U–NU	6	TBMF
	Morelia, Mexico	+	U–N	11	TSDBF
	Careyes, Mexico	+	S–N	9	TSDBF
	Ceiba, Puerto Rico	+	U–NU	8	M

[a]*S* Suburban, *S–N* Suburban–Natural, *S–NU* Suburban–Nonurban, *U–N* Urban–Natural, *U–P* Urban–Peri-urban, *U–NU* Urban–Nonurban

[b]1: Leveau (2013b); 2: Leveau and Leveau (2005); 3: Leveau and Leveau (2006); 4: Juri and Chani (2005); 5: Varela-Contador (2003) 6: Silva et al. (2016); 7: Amâncio et al. (2008); 8: Komar et al. (2012); 9: MacGregor-Fors and Schondube (2012); 10: Rossetti and Giraudo (2003); 11: López-Flores et al. (2009)

[c]*FGS* flooded grasslands and savannas, *M* mangroves, *MFWS* mediterranean forests, woodlands, and scrub, *TBMF* temperate broadleaf and mixed forests, *TGSS* temperate grasslands, savannas, and shrublands, *TSDBF* tropical and subtropical dry broadleaf forests, *TSGSS* tropical and subtropical grasslands, savannas, and shrublands, *TSMBF* tropical and subtropical moist broadleaf forests. Classification based on Olson et al. (2001)

4.3.1.2 Urban Greenspaces

Of the gathered studies, only eight assessed the abundance of bird species within urban greenspaces, of which six were parks (i.e., Cândido 2006; Toledo 2007; Rebolo-Ifran and Fiorini 2010; Ortega-Álvarez and MacGregor-Fors 2010; Carbó-Ramírez and Zuria 2011; González-Oreja et al. 2012), one in a forest remnant

Table 4.3 Bird species showing generally negative abundance responses to urbanization in Latin America

Species	Study location	Response	Contrast[a]	Source[b]	Biome[c]
Southern Lapwing *Vanellus chilensis*	Santa Fé Province, Argentina	+	S–N	1	FGS
	Mar del Plata, Argentina	–	S–NU	2	TGSS
	Santiago de Chile, Chile	+	S	3	MFWS
	Temuco, Chile	–	U–NU	4	MFWS
	Valdivia, Chile	–	U–NU	4	TBMF
	Osorno, Chile	–	U–NU	4	TBMF
White-crested Elaenia *Elaenia albiceps*	Santiago de Chile, Chile	–	S	3	MFWS
	Temuco, Chile	–	U–NU	4	MFWS
	Valdivia, Chile	–	U–NU	4	TBMF
	Osorno, Chile	–	U–NU	4	TBMF
Tropical Kingbird *Tyrannus melancholicus*	Santa Fé Province, Argentina	–	S–N	1	FGS
	Mar del Plata, Argentina	–	S	5	TGSS
	Mar del Plata, Argentina	+	S–NU	2	TGSS
	Buenos Aires Province, Argentina	–	S	6	TGSS
	Careyes, Mexico	–	S–N	7	TSDBF
Austral Thrush *Turdus falcklandii*	Santiago de Chile, Chile	Null	S	3	MFWS
	Temuco, Chile	–	U–NU	4	MFWS
	Valdivia, Chile	–	U–NU	4	TBMF
	Osorno, Chile	–	U–NU	4	TBMF
Grassland Yellow-Finch *Sicalis luteola*	Mar del Plata, Argentina	–	S–NU	2	TGSS
	Temuco, Chile	–	U–NU	4	MFWS
	Valdivia, Chile	–	U–NU	4	TBMF
	Osorno, Chile	–	U–NU	4	TBMF

(continued)

Table 4.3 (continued)

Species	Study location	Response	Contrast[a]	Source[b]	Biome[c]
Rufous-collared Sparrow *Zonotrichia capensis*	Santa Fé Province, Argentina	+	S–N	1	MFWS
	Tafi Viejo, Argentina	–	S–E	8	TSMBF
	Bahia Blanca, Argentina	–	S–NU	9	TGSS
	Mar del Plata, Argentina	–	S	5	TGSS
	Mar del Plata, Argentina	–	S–NU	2	TGSS
	Buenos Aires Province, Argentina	Null	S	6	TGSS
	San Miguel de Tucumán, Argentina	–	U–P	10	TSMBF
	Temuco, Chile	–	U–NU	4	MFWS
	Valdivia, Chile	–	U–NU	4	TBMF
	Osorno, Chile	–	U–NU	4	TBMF
	Santiago de Chile, Chile	Null	S	3	MFWS

[a]*S* Suburban, *S–E* Suburban–Exurban, *S–N* Suburban–Natural, *S–NU* Suburban–Nonurban, *U–P* Urban–Peri-urban, *U–NU* Urban–Nonurban

[b]1: Rossetti and Giraudo (2003); 2: Leveau and Leveau (2005); 3: Varela-Contador (2003); 4: Silva et al. (2016); 5: Leveau (2013b); 6: Leveau and Leveau (2006); 7: MacGregor-Fors and Schondube (2012); 8: Navarro and Antelo (2014); 9: Laiolo (2011); 10: Juri and Chani (2005)

[c]*FGS* flooded grasslands and savannas, *MFWS* mediterranean forests, woodlands, and scrub, *TBMF* temperate broadleaf and mixed forests, *TGSS* temperate grasslands, savannas, and shrublands, *TSDBF* tropical and subtropical dry broadleaf forests, *TSMBF* tropical and subtropical moist broadleaf forests. Classification based on Olson et al. (2001)

(i.e., Suarez-Rubio and Thomlinson 2009), and another in a garden (i.e., González-García 2014). The main goals of these studies were to assess the effect of patch size, different patch traits, and landscape structure. For example, González-Oreja et al. (2012) assessed the response of bird species to vegetation composition and heterogeneity in urban parks of Puebla, Mexico, finding that the abundance of the Vermilion Flycatcher (*Pyrocephalus rubinus*), Bronzed Cowbird (*Molothrus aeneus*), and Canyon Towhee (*Melozone fusca*) positively correlated with native vegetation richness and structural diversity, whereas Rock Pigeon and House Sparrow numbers positively correlated with exotic vegetation richness and structural diversity. Carbó-Ramírez and Zuria (2011) analyzed bird communities within different types of small urban greenspaces in Pachuca (Mexico), finding that bird abundance was lower in median strips when compared to parks and gardens. They also found that the abundance of granivores species (e.g., Rock Pigeon; Inca Dove—*Columbina inca*; House Sparrow; Brown-headed Cowbird—*Molothrus ater*) increased in greenspaces surrounded by adjacent urban matrices with high built cover. Finally, Cândido (2006) found that the Eared Dove, Ruddy Ground-Dove (*Columbina talpacoti*), and Pale-vented Pigeon (*Patagioenas cayennensis*) had higher densities in urban parks at downtown of Uberlândia (Brasil), whereas the Picazuro Pigeon (*Patagioenas picazuro*) and Scaled Dove (*Columbina squamata*) were more common in 'peri-urban' parks.

4.3.2 Demography

Few studies have focused on demographic aspects of bird populations in urban areas of Latin America. Again, most studies are from Brazil, Mexico, and Argentina, and focus on two main topics: breeding biology and mortality (studies are also available for Costa Rica, Colombia, Chile, Cuba, Curacao, El Salvador, and Uruguay). Avian population survival has been analyzed in highly urbanized islands and continental areas. For example, Montambault (2007) surveyed landbirds in the Middle Caicos, Turks, and Caicos Islands with a focus on the presence and territory size of the Bahama Mockingbird (*Mimus gundlachii*). This study followed a population viability analysis to assess its population sustainability using survival and reproductive parameters from other species in different areas. Their results show that under the current urbanization rates, the mockingbird population has a 100% island-wide extinction risk with an average time to extinction of 37 years. Rebolo-Ifran et al. (2015) measured feather corticosterone levels and annual survival in urban and nonurban Burrowing Owls (*Athene cunicularia*) in Argentina. This study focused on data of 183 individuals from 2006 to 2012, and found no change in breeding habitat from urban to nonurban individuals, as well as no differences on survival rates between males and females. However, survival rates were twice as high for the urban individuals when compared to nonurban ones due to lower predation risk occurring in urban areas, which could have contributed to the increased breeding densities currently found in urban Buenos Aires. Moreover, this study did not find a relationship between survival and corticosterone levels, which led to the conclusion that urban life does not increase the physiological stress experienced by individuals of this species, supporting the hypothesis that individuals select breeding habitats according to their tolerance to human disturbance. For other species, higher concentrations of corticosterone and immunoglobulin have been found in urban areas, indicating that many birds are under chronic stress and may be physiologically vulnerable (Chávez-Zichinelli et al. 2013).

Among demographic studies, those focused on the nesting success of birds in urban Latin America are populating the literature. Descriptions of bird breeding biology in urban areas include groups of species (e.g., Mella and Loutit 2007; Tampson and Petry 2008; Zuria and Rendón-Hernández 2010) as well as particular species of different avian groups (Table 4.4). Studies analyzing the breeding success of birds in urban areas of Latin America are less common. For example, in a moderately urbanized area of Brazil, Davanço et al. (2013) studied the reproductive biology of the Pale-breasted Thrush (*Turdus leucomelas*). During three breeding seasons, they monitored 50 nests and recorded an overall nesting success of 57%, which is higher than the results for other thrush populations in the forests of North America (Knupp et al. 1977). Similarly, Carbó-Ramírez et al. (2015) studied

Table 4.4 Studies focused on the bird breeding biology of landbirds in urban Latin America

Species	References
Eared Dove *Zenaida auriculata*	Mella and Loutit (2007)
Inca Dove *Columbina inca*	Zuria and Rendón-Hernández (2010)
Kelp Gull *Larus dominicanus*	Branco et al. (2008)
Short-tailed Hawk *Buteo brachyurus*	Betto-Monsalvo (2012)
Plumbeous Kite *Ictinia plumbea*	Loures-Ribeiro et al. (2003)
Barn Owl *Tyto alba*	Debrot et al. (2001)
Red-crowned Woodpecker *Melanerpes rubricapillus*	Sandoval (2009)
Golden-naped Woodpecker *Melanerpes chrysauchen*	Sandoval (2009)
Yellow-headed Caracara *Milvago chimachima*	De La Ossa and De La Ossa-Lacayo (2011)
Monk Parakeet *Myiopsitta monachus*	Volpe and Aramburú (2011)
Burrowing Parakeet *Cyanoliseus patagonus*	Tella et al. (2014)
Rufous Hornero *Furnarius rufus*	Marreis and Sander (2006)
Masked Water-Tyrant *Fluvicola nengeta*	Fernandes et al. (2007)
Vermilion Flycatcher *Pyrocephalus rubinus*	Zuria and Rendón-Hernández (2010)
White-crested Elaenia *Elaenia albiceps*	Mella and Loutit (2007)
Cave Swallow *Petrochelidon fulva*	Peris and Llanes (1998)
House Wren *Troglodytes aedon*	Zaldúa et al. (2013)
Cactus Wren *Campylorhynchus brunneicapillus*	Zuria and Rendón-Hernández (2010)
Pale-breasted Thrush *Turdus leucomelas*	Davanço et al. (2013)
Austral Thrush *Turdus falcklandii*	Mella and Loutit (2007)
Rufous-backed Robin *Turdus rufopalliatus*	Pineda-López and Malagamba-Rubio (2009), Carbó-Ramírez et al. (2015)
Curve-billed Thrasher *Toxostoma curvirostre*	Zuria and Rendón-Hernández (2010)

(continued)

Table 4.4 (continued)

Species	References
European Starling *Sturnus vulgaris*	Gómez-Aíza and Zuria (2012)
House Finch *Haemorhous mexicanus*	Zuria and Rendón-Hernández (2010)
Rufous-collared Sparrow *Zonotrichia capensis*	Mella and Loutit (2007)
Canyon Towhee *Melozone fusca*	Torres-Torres and Salgado-Ortiz (2015), Zuria and Rendón-Hernández (2010)

the breeding success of eight nests of the Rufous-backed Robin (*Turdus rufopalliatus*) in an urban area of central Mexico. On average, 1.88 ± 0.35 nestlings hatched per nest, and 1.25 ± 0.45 fledglings/nest left the nest. In Argentina, nesting sites of the Cliff-nesting Burrowing Parrot (*Cyanoliseus patagonus*) were studied in the urbanized area of Bahía Blanca. Most colonies (68%) were found in human-made structures, suggesting that anthropogenic substrates have allowed population expansion (Tella et al. 2014; Chap. 5). Pretelli et al. (2015) studied the effects of fragmentation on the reproduction of two species of grassland birds, the Speckled Tyrant (*Hymenops perspicillatus*) and Brown-and-yellow Marshbird (*Pseudoleistes virescens*), in grassland patches embedded within urban and agricultural landscape matrices in Argentina. In this study, cumulative probability of nest survival was higher in urban patches (0.36) than in agricultural ones (0.20), suggesting that some predators in urban landscapes can specialize in alternative food sources such as anthropogenic food, reducing predation pressure on bird nests.

Nest parasitism has been rarely studied in urban areas of Latin America. Torres-Torres and Salgado-Ortiz (2015) analyzed nest parasitism by Bronzed Cowbirds in 'urban' and 'suburban' areas of the city of Morelia (Mexico). They found a parasite prevalence of ~68% for the Canyon Towhee (*Melozone fusca*), which was higher in 'urban' than in 'suburban areas'. Although nest predation has been also rarely studied in urban areas of Latin America, experiments using artificial nests have been conducted in some urbanized environments (e.g., Cervantes-Cornhis et al. 2009; López-Flores et al. 2009; de Almeida e Silva and Voltolini 2015; Rivera-López and MacGregor-Fors 2016). In the city of Morelia (Mexico), artificial nest predation was analyzed along an urban gradient, and nest survivorship was found to be higher in shrublands than in urbanized environments (López-Flores et al. 2009). Also in Mexico, Rivera-López and MacGregor-Fors (2016) assessed the degree of artificial nest predation pressure relative to spatial and habitat traits in the neotropical city of Xalapa, finding higher predation rates in 'peri-urban' areas, and concluding that artificial nest predation was driven by the interaction between geographical location and urbanization intensity.

4.4 Opportunities and Future Research Needs

After reviewing the literature focused on avian abundance and demography in urban Latin America, we found a clear lack of studies in highly biodiverse countries, such as Ecuador, Venezuela, Colombia, as well as countries of Central America, leaving an important information gap. Also, information from countries where urban bird populations have been studied, such as Brazil, Mexico, Argentina, and Chile, came mainly from comparisons along urban matrices, while data of species abundances and/or demographical parameters from urban greenspaces were scarcer. Considering that these areas can harbor high densities of bird species that are less common in other urban habitats (Carbó-Ramírez and Zuria 2011; Leveau and Leveau 2016), there is a need to focus research efforts on species numbers in greenspaces. Another important topic is the temporal variation of bird populations in urban areas. This is relevant because we may estimate long-term declines of bird species or the spread of invasive species (Chap. 9).

A better understanding of habitat use by species is achieved by considering multiple spatial and temporal scales. Unfortunately, this kind of information is scarce in the urban Latin American literature (Suarez-Rubio and Thomlinson 2009; Rosselli and Stiles 2012). For example, local traits as well as landscape composition and structure are determinants of species presence in urban greenspaces (Taylor et al. 2016). Additionally, we need more studies focusing on which bird traits allow species to thrive in urban environments. Although we know that omnivorous species that feed on the ground gregariously are more abundant in highly urbanized areas (Ortega-Álvarez and MacGregor-Fors 2009; Conole and Kirkpatrick 2011; Leveau 2013a; Silva et al. 2016), the effect of urbanization on other traits such as nectarivory and frugivory may vary according to the biome where the cities are located (Reynaud and Thioulouse 2000; Leveau and Leveau 2004).

Information regarding the demographical parameters of urban bird populations (e.g., survival, breeding success, mortality) is even less common for urban Latin America. In this sense, there is an urgent need for detailed studies that identify the main predictors of the reproductive success of native bird populations within urban areas (e.g, habitat connectivity, nest predation in urban ecotones, parasitism, diseases, competition with exotic species). Such information would set the grounds for urban managers and planners to promote the conditions for urban birds in Latin America.

It is interesting to note that we found no studies considering the effects of urbanization on population connectivity and metapopulation structure. These approaches are crucial if we aim to manage and conserve bird populations in cities. Urban areas include few, if existent, remnant patches of natural habitat for wildlife immersed within the urban matrix, and many bird species are not able to thrive through developed areas (Tremblay and St. Clair 2011; Delaney 2014). When connectivity and gene flow between populations is restricted, problems such as inbreeding, loss of genetic diversity, and local extinction may arise (Reed 2004). Research suggests that regional populations can persist in fragmented landscapes if they act as a metapopulation, where subpopulations fluctuate independently (i.e., asynchronic), with the possibility of recolonization to avoid that any of the subpopulations becomes extinct (Hanski and Gilpin 1991; Koenig 1998; Padilla and Rodewald 2015). It is also imperative to know the requirements for dispersal within and throughout cities, including migratory species (as assessed by MacGregor-Fors et al. 2010a), in order to enforce effective management practices.

Increasing our knowledge about bird abundance and demography in urban Latin America could aid decision-making for developers, managers, and planners. Preserving viable bird populations in urban areas will decrease the probability of regional extinction of those species able to thrive within cities, enhancing the aesthetic value of urban greenspaces, and promoting public awareness about biodiversity (Savard et al. 2000; Fernández Juricic 2004; Carbó-Ramírez and Zuria 2011).

Acknowledgments IZ thanks CONACyT through the thematic network REFAMA (251272) "Red temática Biología, manejo y conservación de la fauna nativa en ambientes antropizados". LML thanks Lucia Gonzalez Salinas for her love and patience.

References

Amâncio S, Barbosa de Souza V, Melo C (2008) *Columba livia* e *Pitangus sulphuratus* como indicadoras de qualidade ambiental em área urbana. Rev Bras Ornithol 16:32–37

Amrhein V (2014) Wild bird feeding (probably) affects avian urban ecology. In: Gil D, Brumm H (eds) Avian urban ecology. Oxford University Press, Oxford, pp 29–37

Arnold TW, Zink RM (2011) Collision mortality has no discernible effect on population trends of North American birds. PLoS One 6:e24708

Aronson MFJ, La Sorte FA, Nilon CH et al (2014) A global analysis of the impacts of urbanization on bird and plant diversity reveals key anthropogenic drivers. Proc R Soc Lond B Biol Sci 281:20133330

Ausprey IJ, Rodewald AD (2011) Postfledging survivorship and habitat selection across a rural-to-urban landscape gradient. Auk 128:293–302

Baker PJ, Molony SE, Stone E et al (2008) Cats about town: is predation by free-ranging pet cats *Felis catus* likely to affect urban bird populations? Ibis 150:86–99

Barbar F, Werenkraut V, Morales JM et al (2015) Emerging ecosystems change the spatial distribution of top carnivores even in poorly populated areas. PLoS One 10:e0118851

Beissinger SR, Osborne DR (1982) Effects of urbanization on avian community organization. Condor 84:75–83

Bellocq MI, Filloy J, Garaffa PI (2008) Influence of agricultural intensity and urbanization on the abundance of the raptor chimango caracara (*Milvago chimango*) in the Pampean region of Argentina. Ann Zool Fenn 45:128–134

Betto-Monsalvo JA (2012) Reproduçao de *Buteo brachyurus* em um parque urbano de São Paulo, sudeste do Brasil. Atual Ornithol 170:33–40

Blair RB (1996) Land use and avian species diversity along an urban gradient. Ecol Appl 6:506–519

Blair R (2004) The effects of urban sprawl on birds at multiple levels of biological organization. Ecol Soc 9:2

Blair RB, Johnson EM (2008) Suburban habitats and their role for birds in the urban-rural habitat network: points of local invasion and extinction? Landsc Ecol 23:1157–1169

Bock CE, Jones ZF (2004) Avian habitat evaluation: should counting birds count? Front Ecol Environ 2:403–410

Bonnington C, Gaston KJ, Evans KL (2013) Fearing the feline: domestic cats reduce avian fecundity through trait-mediated indirect effects that increase nest predation by other species. J Appl Ecol 50:15–24

Branco JO, de Azevedo Júnior MS et al (2008) Reprodução de *Larus dominicanus* (Aves, Laridae) em ambiente urbano. Rev Bras Ornithol 16:240–242

Campbell MON (2014) The impact of urbanization and agricultural development on vultures in El Salvador. Vulture News 66:16–28

Cândido CR (2006) Distribução de espécies de columbídeos (Aves: Columbiformes), na zona urbana de Uberlândia, MG. MSc Thesis, Universidade Nacional de Uberlândia, Brasil

Carbó-Ramírez P, Zuria I (2011) The value of small urban greenspaces for birds in a Mexican city. Landsc Urban Plan 100:213–222

Carbó-Ramírez P, González-Arrieta RA, Zuria I (2015) Breeding biology of the Rufous-backed Robin (*Turdus rufopalliatus*) in an urban area outside its original distribution range. Wilson J Ornithol 127:515–521

Carvalho CEA, Marini MÂ (2007) Distribution patterns of diurnal raptors in open and forested habitats in south-eastern Brazil and the effects of urbanization. Bird Conserv Int 17:367–380

Catterall C (2004) Birds, garden plants and suburban bushlots: where good intentions meet unexpected outcomes. In: Lunney D, Burgin S (eds) Urban wildlife: more than meets the eye. Royal Zoological Society of New South Wales, Mosman, pp 22–31

Caula S, Marty P, Martin JL (2008) Seasonal variation in species composition of an urban bird community in Mediterranean France. Landsc Urban Plan 87:1–9

Cervantes-Cornhis E, Zuria I, Castellanos I (2009) Depredación de nidos artificiales en cercas vivas de un sistema agro-urbano en Hidalgo, México. Interciencia 34:777–783

Chace JF, Walsh JJ (2006) Urban effects on native avifauna: a review. Landsc Urban Plan 74:46–69

Chávez-Zichinelli CA, Macgregor-Fors I, Quesada J et al (2013) How stressed are birds in an urbanizing landscape? Relationships between the physiology of birds and three levels of habitat alteration. Condor 115:84–92

Clergeau P, Croci S, Jokimäki J et al (2006) Avifauna homogenisation by urbanisation: analysis at different European latitudes. Biol Conserv 127:336–344

Conole LE, Kirkpatrick JB (2011) Functional and spatial differentiation of urban bird assemblages at the landscape scale. Landsc Urban Plan 100:11–23

Conway CJ, Garcia V, Smith MD et al (2006) Comparative demography of Burrowing Owls in agricultural and urban landscapes in southeastern Washington. J Field Ornithol 77:280–290

Croci S, Butet A, Clergeau P (2008) Does urbanization filter birds on the basis of their biological traits. Condor 110:223–240

Davanço PV, de Oliveira LS, Sousa LMS et al (2013) Breeding life-history traits of the Pale-breasted Thrush (*Turdus leucomelas*) in southeastern Brazil. Ornithol Neotrop 24:401–411

Debrot AO, De Freitas JA, Brouwer A et al (2001) The Curaçao Barn Owl: status and diet, 1987–1989. Caribb J Sci 37:185–193

de Almeida e Silva AS, Voltolini JC (2015) Predação de ninhos artificiais em dois fragmentos urbanos de Mata Atlântica no sudeste do Brasil. Rev Biociênc 21:29–37

De La Ossa J, De La Ossa-Lacayo A (2011) Aspectos de la densidad poblacional e historia natural de *Milvago chimachima* (Aves: Falconidae) en el área urbana de Sincelejo (Sucre, Colombia). Univ Sci 16:63–69

Delaney KS (2014) Landscape genetics of urban bird populations. In: Gil D, Brumm H (eds) Avian urban ecology. Oxford University Press, Oxford, pp 143–154

Dunn JC, Hamer KC, Benton TG (2010) Fear for the family has negative consequences: indirect effects of nest predators on chick growth in a farmland bird. J Appl Ecol 4:994–1002

Evans BS, Ryder TB, Reitsma R et al (2015) Characterizing avian survival along a rural-to-urban land use gradient. Ecology 96:1631–1640

Evans KL, Chamberlain DE, Hatchwell BJ et al (2011) What makes an urban bird? Glob Chang Biol 17:32–44

Fernandes VM, Tomaz VC, Alves MAS (2007) Reprodução da Lavadeira-mascarada *Fluvicola nengeta* (Aves: Tyrannidae) em área urbana no Rio de Janeiro, RJ. Anais do VIII Congesso de Ecologia do Brasil, Caxambu, 23–28 September 2007
Fernández Juricic E (2004) Spatial and temporal analysis of the distribution of forest specialists in an urban-fragmented landscape (Madrid, Spain): implications for local and regional bird conservation. Landsc Urban Plan 69:17–32
Fischer JD, Cleeton SH, Lyons TP et al (2012) Urbanization and the predation paradox: the role of trophic dynamics in structuring vertebrate communities. Bioscience 62:809–818
Fischer JD, Schneider SC, Ahlers AA et al (2015) Categorizing wildlife responses to urbanization and conservation implications of terminology. Conserv Biol 29:1246–1250
Germaine SS, Rosenstock SS, Schweinsburg RE et al (1998) Relationships among breeding birds, habitat, and residential development in Greater Tucson, Arizona. Ecol Appl 8:680–691
Gómez-Aíza L, Zuria I (2012) Registros del estornino pinto (*Sturnus vulgaris*) en la ciudad de Pachuca, Hidalgo y evidencias de actividad reproductiva. Huitzil 13:146–150
González-García A (2014) Estudio ecológico de los patios urbanos en León (Nicaragua): Su contribución a la biodiversidad, el paisaje y el patrimonio. PhD Dissertation, Universidad de Alcalá
González-Oreja JA, Barillas-Gómez AL, Bonache-Regidor C et al (2012) Does habitat heterogeneity affect bird community structure in urban parks? In: Lepczyk CA, Warren PS (eds) Urban bird ecology and conservation. Studies in avian biology. University of California Press, Berkeley, pp 1–16
Gowaty PA (1984) House sparrows kill eastern bluebirds. J Field Ornithol 55:378–380
Hager SB, Trudell H, McKay KJ et al (2008) Bird density and mortality at windows. Wilson J Ornithol 120:550–564
Hanski I, Gilpin M (1991) Metapopulation dynamics: brief history and conceptual domain. Biol J Linn Soc 42:3–16
Horak P, Lebreton JD (1998) Survival of adult Great Tits *Parus major* in relation to sex and habitat; a comparison of urban and rural populations. Ibis 140:205–209
Ikin K, Barton PS, Knight E et al (2014) Bird community responses to the edge between suburbs and reserves. Oecologia 174:545–557
Juri MD, Chani JM (2005) Variación en la composición de comunidades de aves a lo largo de un gradiente urbano (Tucumán, Argentina). Acta Zool Lilloana 49:49–57
Kark S, Iwaniuk A, Schalimtzek A et al (2007) Living in the city: can anyone become an 'urban exploiter'? J Biogeogr 34:638–651
Kennedy CM, Marra PP, Fagan WF et al (2010) Landscape matrix and species traits mediate responses of Neotropical resident birds to forest fragmentation in Jamaica. Ecol Monogr 80:651–669
Klem D Jr (1990) Collisions between birds and windows: mortality and prevention. J Field Ornithol 61:120–128
Knupp DM, Owen RB, Dimond JB (1977) Reproductive biology of American Robins in northern Maine. Auk 94:80–85
Koenig WD (1998) Spatial autocorrelation in California land birds. Conserv Biol 12:612–660
Komar N, Bessoff K, Diaz A et al (2012) Avian hosts of West Nile virus in Puerto Rico. Vector Borne Zoonotic Dis 12:47–54
Laiolo P (2011) Homogenisation of birdsong along a natural–urban gradient in Argentina. Ethol Ecol Evol 23:274–287
Leston LFV, Rodewald AD (2006) Are urban forests ecological traps for understory birds? An examination using northern cardinals. Biol Conserv 131:566–574
Leveau LM (2013a) Bird traits in urban-rural gradients: how many functional groups are there? J Ornithol 154:655–662
Leveau LM (2013b) Relaciones aves-hábitat en el sector suburbano de Mar del Plata, Argentina. Ornithol Neotrop 24:201–212
Leveau CM, Leveau LM (2005) Avian community response to urbanization in the Pampean region, Argentina. Ornithol Neotrop 16:503–510

Leveau CM, Leveau LM (2006) Ensambles de aves en calles arboladas de tres ciudades costeras del sudeste de la provincia de Buenos Aires, Argentina. El Hornero (B. Aires) 21:25–30

Leveau LM, Leveau CM (2004) Comunidades de aves en un gradiente urbano de la ciudad de Mar del Plata Argentina. El Hornero (B. Aires) 19:13–21

Leveau LM, Leveau CM (2016) Does urbanization affect the seasonal dynamics of bird communities in urban parks? Urban Ecosyst 19:631–647

López-Flores V, MacGregor-Fors I, Schondube JE (2009) Artificial nest predation along a Neotropical urban gradient. Landsc Urban Plan 92:90–95

Loss SR, Will T, Marra PP (2012) Direct human-caused mortality of birds: improving quantification of magnitude and assessment of population impacts. Front Ecol Environ 10:357–364

Loss SR, Will T, Marra PP (2014a) Estimation of bird-vehicle collision mortality on U.S. roads. J Wildl Manag 78:763–771

Loss SR, Will T, Marra PP (2014b) Bird-building collisions in the United States: estimates of annual mortality and species vulnerability. Condor 116:8–23

Loss SR, Will T, Marra PP (2015) Direct mortality of birds from anthropogenic causes. Annu Rev Ecol Evol Syst 46:99–120

Loures-Ribeiro A, Gimenes MR, dos Anjos L (2003) Observações sobre o comportamento reprodutivo de *Ictinia plumbea* (Falconiformes: Accipitridae) no Campus da Universidade Estadual de Maringá, Paraná, Brasil. Arajuba 11:85–87

Lozano J, Malo AF (2013) Relationships between human activity and richness and abundance of some bird species in the Paraguay River (Pantanal, Brazil). Ardeola 60:99–112

MacGregor-Fors I (2011) Misconceptions or misunderstandings? On the standardization of basic terms and definitions in urban ecology. Landsc Urban Plan 100:347–349

MacGregor-Fors I, Schondube JE (2012) Urbanizing the wild: shifts in bird communities associated to small human settlements. Rev Mex Biodivers 83:477–486

MacGregor-Fors I, Morales-Pérez L, Schondube JE (2010a) Migrating to the city: responses of Neotropical migrant bird communities to urbanization. Condor 112:711–717

MacGregor-Fors I, Morales-Pérez L, Quesada J et al (2010b) Relationship between the presence of House Sparrows (*Passer domesticus*) and Neotropical bird community structure and diversity. Biol Invasions 12:87–96

Mannan RW, Estes WA, Matter WJ (2004) Movements and survival of fledgling Cooper's Hawks in an urban environment. J Raptor Res 38:26–34

Marreis IT, Sander M (2006) Preferência ocupacional de ninhos de João-de-barro (*Furnarius rufus*, Gmelin) entre área urbanizada e natural. Biodivers Pampeana 4:29–31

Martin TE, Lloyd P, Bosque C et al (2011) Growth rate variation among passerine species in tropical and temperate sites: an antagonistic interaction between parental food provisioning and nest predation risk. Evolution 65:1607–1622

Martin LB, Boruta M (2014) The impacts of urbanization on avian disease transmission and emergence. In: Gil D, Brumm H (eds) Avian urban ecology. Oxford University Press, Oxford, pp 116–143

Marzluff JM, Bowman R, Donnelly R (2001) A historical perspective on urban bird research: trends, terms, and approaches. In: Marzluff JM, Bowman R, Donnelly R (eds) Avian ecology and conservation in an urbanizing world. Kluwer Academic Publishers, New York, pp 1–17

McKinney ML (2002) Urbanization, biodiversity, and conservation. Bioscience 52:883–890

Mella JE, Loutit A (2007) Ecología comunitaria y reproductiva de aves en cerros islas y parques de Santiago. Bol Chil Ornithol 13:13–27

Merenlender AM, Reed SE, Heise KL (2009) Exurban development influences woodland bird composition. Landsc Urban Plan 92:255–263

Montague-Drake RM, Lindenmayer DB, Cunningham RB et al (2011) A reverse keystone species affects the landscape distribution of woodland avifauna: a case study using the Noisy Miner (*Manorina melanocephala*) and other Australian birds. Landsc Ecol 26:1383–1394

Montambault JR (2007) Conservation status and potential of West Indian endemic bird species in a rapidly suburbanizing landscape, Middle Caicos, Turks & Caicos Islands. PhD Dissertation, University of Florida

Navarro CI, Antelo CM (2014) Comunidades de aves presentes en un área residencial suburbana y en un área exurbana en el Departamento Tafí Viejo (Tucumán, Argentina). Acta Zool Lilloana 58:133–146

Newton I (1998) Population limitation in birds. Academic Press, London

Olson DM, Dinerstein E, Wikramanayake ED, Burgess ND, Powell GVN, Underwood EC, D'amico JA, Itoua I, Strand HE, Morrison JC, Loucks CJ, Allnutt TF, Ricketts TH, Kura Y, Lamoreux JF, Wettengel WW, Hedao P, Kassem KR (2001) Terrestrial ecoregions of the world: A new map of life on earth. Bioscience 51:933–938

Ortega-Álvarez R, MacGregor-Fors I (2009) Living in the big city: effects of urban land-use on bird community structure, diversity, and composition. Landsc Urban Plan 90:189–195

Ortega-Álvarez R, MacGregor-Fors I (2010) What matters most? Relative effect of urban habitat traits and hazards on urban park birds. Ornithol Neotrop 21:519–533

Ortega-Álvarez R, MacGregor-Fors I (2011) Dusting-off the file: a review of knowledge on urban ornithology in Latin America. Landsc Urban Plan 101:1–10

Padilla BJ, Rodewald AD (2015) Avian metapopulation dynamics in a fragmented urbanizing landscape. Urban Ecosyst 18:239–250

Parsons H, Major RE, French K (2006) Species interactions and habitat associations of birds inhabiting urban areas of Sydney, Australia. Austral Ecol 31:217–227

Peris S, Llanes A (1998) Breeding observations of Cave Swallows (*Hirundo fulva*) at two Cuban sites. Ornithol Neotrop 9:209–2012

Phillips J, Nol E, Burke D et al (2005) Impacts of housing development on Wood Thrush nesting success in hardwood forest fragments. Condor 107:97–106

Pineda-López R, Malagamba-Rubio A (2009) Primeros registros de presencia y reproducción del mirlo dorso rufo (*Turdus rufopalliatus*) en la ciudad de Querétaro, Querétaro, México. Huitzil 10:66–70

Pretelli MG, Isacch JP, Cardoni DA (2015) Effects of fragmentation and landscape matrix on the nesting success of grassland birds in the Pampas grasslands of Argentina. Ibis 157:688–699

Rebolo-Ifran NR, Fiorini VD (2010) European Starling (*Sturnus vulgaris*): population density and interactions with native species in Buenos Aires urban parks. Ornithol Neotrop 21:507–518

Rebolo-Ifran N, Carrete M, Sanz-Aguilar A et al (2015) Links between fear of humans, stress and survival support a non-random distribution of birds among urban and rural habitats. Sci Rep 5:13723

Reed DH (2004) Extinction risk in fragmented habitats. Anim Conserv 7:181–191

Reynaud PA, Thioioulouse J (2000) Identification of birds as biological markers along a Neotropical urban-rural gradient (Cayenne, French Guiana), using co-inertia analysis. J Environ Manag 59:121–140

Rivera-López A, MacGregor-Fors I (2016) Urban predation: a case study assessing artificial nest survival in a Neotropical city. Urban Ecosyst 19:649–655

Robb GN, McDonald RA, Chamberlain DE et al (2008) Food for thought: supplementary feeding as a driver of ecological change in avian populations. Front Ecol Environ 6:476–484

Rodewald AD, Gehrt SD (2014) Wildlife population dynamics in urban landscapes. In: RA MC, Moorman CE, Peterson MN (eds) Urban wildlife conservation. Springer, New York, pp 117–147

Rodewald AD, Shustack DP (2008a) Urban flight: understanding individual and population-level responses of Neactic-Neotropical migratory birds to urbanization. J Anim Ecol 77:83–91

Rodewald AD, Shustack DP (2008b) Consumer resource matching in urbanizing landscapes: are synanthropic species over-matching. Ecology 89:515–521

Rosselli L, Stiles FG (2012) Local and landscape environmental factors are important for the conservation of endangered wetland birds in a high Andean plateau. Waterbirds 35:453–469

Rossetti MA, Giraudo AR (2003) Comunidades de aves de bosques fluviales habitados y no habitados por el hombre en el río Paraná medio Argentina. El Hornero (B. Aires) 18:89–96

Sandoval L (2009) Densidad de sitios para anidar y su uso por dos especies de pájaro carpintero (*Melanerpes rubricapillus* y *M. chrysauchen*) (Piciformes: Picidae) en un gradiente urbano de Costa Rica. Rev Biol Trop 57:351–355

Savard JPL, Clergeau P, Mennechez G (2000) Biodiversity concepts and urban ecosystems. Landsc Urban Plan 48:131–142

Shanahan DF, Strohbach MW, Warren PS et al (2014) The challenges of urban living. In: Gil D, Brumm H (eds) Avian urban ecology. Oxford University Press, Oxford, pp 3–20

Shochat E, Warren PS, Faeth SH et al (2006) From patterns to emerging processes in mechanistic urban ecology. Trends Ecol Evol 21:186–191

Silva CP, Sepúlveda RD, Barbosa O (2016) Nonrandom filtering effect on birds: species and guilds response to urbanization. Ecol Evol 6:3711–3720

Suarez-Rubio M, Thomlinson JR (2009) Landscape and patch–level factors influence bird communities in an urbanized tropical island. Biol Conserv 142:1311–1321

Sumasgutner P, Nemeth E, Tebb G et al (2014) Hard times in the city-attractive nest sites but insufficient food supply lead to low reproduction rates in a bird of prey. Front Zool 11:48

Tampson VE, Petry MV (2008) Nidificação e análise das guildas alimentares de aves no Morro do Espelho, na zona urbana de São Leopoldo-RS. Biodivers Pampeana 6:63–69

Taylor JJ, Lepczyk CA, Brown DG (2016) Patch and matrix level influences on forest birds at the rural–urban interface. Landsc Ecol 31:1005–1020

Tella JL, Canale A, Carrete M et al (2014) Anthropogenic nesting sites allow urban breeding in burrowing parrots *Cyanoliseus patagonus*. Ardeola 61:344–321

Thornton DH, Branch LC, Sunquist ME (2012) Response of large galliforms and tinamous (Cracidae, Phasianidae, Tinamidae) to habitat loss and fragmentation in northern Guatemala. Oryx 46:567–576

Toledo MCB (2007) Análise das áreas verdes urbanas em diferentes escalas visando a conservação da avifauna. PhD Dissertation, Universidad Estadual Paulista, Brasil

Torres-Torres MT, Salgado-Ortiz J (2015) Efecto del parasitismo de nidos por el Tordo Ojirojo (*Molothrus aeneus*) sobre el éxito reproductivo de poblaciones urbanas y suburbanas del Rascador pardo (*Melozone fusca*) en Morelia Michoacán. Biológicas 17:19–25

Tremblay MA, St. Clair CC (2011) Permeability of a heterogeneous urban landscape to the movements of forest songbirds. J Appl Ecol 48:679–688

van Heezik Y, Smyth A, Adams A et al (2010) Do domestic cats impose an unsustainable harvest on urban bird populations? Biol Conserv 143:121–130

Van Horne B (1983) Density as a misleading indicator of habitat quality. J Wildl Manag 47:893–901

Varela-Contador SI (2003) Calidad de la vegetación urbana como hábitat para aves. El caso de Santiago de Chile. BSc Thesis, Universidad de Chile

Villaneda Rey M, Rosselli L (2011) Abundancia del Chamón Parásito (*Molothrus bonariensis*, Icteridae) en 19 humedales de la Sabana de Bogotá, Colombia. Ornitol Colomb 11:37–48

Volpe NL, Aramburú RM (2011) Preferencias de nidificación de la cotorra argentina (*Myiopsitta monachus*) en un área urbana de Argentina. Ornithol Neotrop 22:111–119

Weaving MJ, White JG, Isaac B et al (2016) Adaptation to urban environments promotes high reproductive success in the Tawny Frogmouth (*Podargus strigoides*), an endemic nocturnal bird species. Landsc Urban Plan 150:87–95

Whittaker KA, Marzluff JM (2009) Species-specific survival and relative habitat use in an urban landscape during the postfledging period. Auk 126:288–299

Zaldúa N, Hernández D, Rodríguez-Tricot L (2013) Evaluación del uso de cajas nido por parte de aves urbanas. Bol Soc Zool Urug 22:39–45

Zuria I, Rendón-Hernández G (2010) Notes on the breeding biology of common resident birds in an urbanized area of Hidalgo, Mexico. Huitzil 11:35–41

Chapter 5
'Green' or 'Gray'? Infrastructure and Bird Ecology in Urban Latin America

Ana Faggi and Sabina Caula

Abstract Urban systems are highly complex and heterogeneous, comprised basically by two types of infrastructure. On the one hand, 'gray' infrastructure includes all artificial structures and buildings, often impervious, and their components. On the other hand, 'green' infrastructure is conformed by all remaining and planted vegetation within an urban center. In this chapter, we reviewed the global response of birds to these two main types of urban infrastructures, and then focus on urban Latin America. Few studies have focused on urban 'gray' infrastructure, with many of them biased toward its close relationships with exotic, invasive, and/or generalist bird species. On the contrary, many studies have been performed focusing on the 'green' infrastructure of cities. Due to the amount of knowledge regarding 'green' infrastructure, we approached it considering greenspace location, size, heterogeneity, vegetation traits, management, and human activities. Our review shows that there is a pressing need for studies that consider 'gray' comparison baselines when assessing the effect of urbanization on birds. Also, studies need to focus on the 'gray' matrix in which urban greenspaces are embedded, as we have little to null information regarding it, and it is essential to plan healthier and sustainable cities. Another important gap that needs to be bridged is the role that exotic plants have not only on bird ecology, but also on the ecology of cities. Finally, we suggest to include other conditions that, although are not as representative in all urban centers, are highly important for birds and other wildlife groups. Specifically, we consider it

A. Faggi (✉)
Museo Argentino de Ciencias Naturales-Consejo Nacional de Investigaciones Científicas y Técnicas, Universidad de Flores, Facultad de Ingeniería, Gallardo, 470 Buenos Aires, Argentina
e-mail: afaggi2003@yahoo.com.ar

S. Caula
Centro de Investigaciones Ambiente, Biología y Química (AMBIOQUIM), Facultad Experimental de Ciencias y Tecnología, Universidad de Carabobo, Naguanagua, Carabobo, Venezuela

I. MacGregor-Fors, J.F. Escobar-Ibáñez (eds.), *Avian Ecology in Latin American Cityscapes*, DOI 10.1007/978-3-319-63475-3_5

crucial to join efforts on studying water bodies ('blue' infrastructure) and wastelands ('brown' infrastructure) to increase our understanding of the role that these urban components play on bird conservation in urban Latin America.

Keywords Citywide • Gray infrastructure • Green infrastructure • Streetscapes • Urban ecology

5.1 Urban Infrastructure

Urban centers are unique systems within landscapes, characterized by high densities of built-up and hard relatively impervious surfaces (Eldredge and Horenstein 2014). Throughout their territory, cities are comprised by complex components pertaining to different land-uses, creating highly heterogeneous systems (Alberti 2005). In terms of infrastructure, urban centers are mainly comprised by built-up surfaces (referred to as 'gray' infrastructure hereafter) and areas with vegetation (referred to as 'green' infrastructure hereafter). The former includes buildings, thoroughfares, and other types of constructions and related components (e.g., poles, cables, antennas, rods, air-conditioning units). The latter includes the sum of all vegetation components within urban settlements, ranging from large native vegetation patches or remnants to isolated street trees (e.g., parks, private gardens, cemeteries, median-strip trees, agricultural fields; Chamberlain et al. 2004; Andersson et al. 2007; Carbó-Ramírez and Zuria 2011; Strohbach et al. 2013). In this chapter, we first summarize our knowledge on the general relationships between 'gray' and 'green' infrastructures with birds, then focus on the literature for Latin America, and conclude highlighting research gaps and future directions.

5.2 Global Role of 'Gray' and 'Green' Infrastructure on Birds

'Gray' infrastructure is highly diverse and complex, reflecting the history and ongoing cultural and economic processes of a city (Seto et al. 2011). This type of infrastructure can be categorized in low, medium, and high densities in a variety of land-uses (e.g., residential, commercial, industrial; MacGregor-Fors 2011). The proportion of 'gray' infrastructure has been commonly used as a proxy to describe the degree of urban development in urban ecology studies (McDonnell et al. 1997; McKinney 2008; Aronson et al. 2014). In fact, an important number of avian ecology studies in urban areas have focused on urbanization intensity gradients (also referred to as urban–rural, urban–wildland, urban–ex-urban gradients; McDonnell and Hahs 2008).

'Gray' infrastructure generally represents a limitation for birds. As reviewed and discussed by Delgado-V and Santiago-Alarcon (Chap. 7), there are numerous urban-related threats to birds, among which several are related to the built-up structures and their components (e.g., collisions with windows and power lines). Yet, previous literature reviews have shown the complexity of the response of birds throughout urbanization gradients, often driving depauperated communities toward the more urban end, mainly comprised by a set of bird species able to dwell, and even thrive, in urban areas (Croci et al. 2008; McDonnell and Hahs 2008; McKinney 2008; Puga-Caballero et al. 2014). Given the nature of urban areas, with high extensions of impervious surfaces and a different array of components, resources, and hazards, birds have shown important acclimation and adaptive responses in these systems (Emlen 1974; McDonnell and Hahs 2015; Chap. 6). Many examples have been recorded in the literature, such as raptors using artificial substrates for nesting (e.g., buildings, bridges, power lines, transmission and communication towers; Chace and Walsh 2006), hummingbirds nesting on open-sky cables (Escobar-Ibáñez and MacGregor-Fors 2015), and parakeets nesting in air-conditioning units (Souza 2016; Chap. 8).

Several studies have shown how urban birds are related with urban 'gray' infrastructure. In a pioneer study, Emlen (1974) underlined the role of house tops and their components (e.g., air-conditioning systems, antennas) as attractive resting and perching sites. The style, density, height, and heterogeneity of buildings have also been shown to be relevant factors for urban birds, with 'ornate older buildings' providing potential nesting and roosting sites, and dense tall buildings related to higher bird abundances, mainly driven by generalist 'urban-dwelling' species (also referred in the literature as 'urban-exploiters'; DeGraaf and Wentworh 1986; Sacchi et al. 2002; MacGregor-Fors et al. 2012; Pellissier et al. 2012).

Although some bird species can benefit from urban 'gray' infrastructure, the 'green' component is crucial for the establishment and permanence of native bird communities in urban systems due to the wide range of resources they provide (Hadidian et al. 1997; Tratalos et al. 2007; MacGregor-Fors and Schondube 2011). Urban 'green' infrastructure is often embedded within the 'gray' one, being isolated at different spatial scales (Davies and Glick 1978; Fernández-Juricic and Jokimäki 2001). When major greenspaces are not connected among them, there are important components of the urban 'green' infrastructure that play crucial roles as ecological connectors (Sodhi et al. 1999; Sandstrom et al. 2006; Evans et al. 2009; MacGregor-Fors et al. 2009).

Urban 'green' infrastructure is mainly determined by human decisions and includes a wide range of different components, such as urban parks, plazas, gardens, lawns, agricultural space, tree-lined avenues and leafy boulevards, natural ecosystem remnants, nature reserves, and cemeteries (Forrest and Konijnendijk 2005; Nowak and Greenfield 2012; Escobedo et al. 2015; Konijnendijk 2016). With urbanization, preexistent conditions are replaced by artificial structures and a mix of native and exotic vegetation, mainly ornamental, which have great adaptability to urban conditions. Many plant species that grow in reference vegetation communities do not look 'attractive' for people and are often perceived to be associated with disservices (e.g., accidents, damages on the 'gray' infrastructure; Gómez-Baggethun and Barton 2013). In urban Latin America, the main factor determining the distribution of urban vegetation seems to be related with socioeconomic level, being scarcer in the low-income sectors incurring in environmental justice issues (Pedlowski et al. 2002; Escobedo et al. 2006; Vázques-Fuentes 2008).

Several traits of the urban 'green' infrastructure have been related with bird diversity, often focused on greenspaces, although few studies focus on street trees (e.g., Donovan and Butry 2010; Carbó-Ramírez and Zuria 2011; Avolio et al. 2015). Despite the findings of different studies are diverse in the type, strength, and direction of such relationships, many vegetation traits have been shown to drive the diversity and composition of urban birds, among which greenspace size, vegetation complexity, and species origin head the list (Nuorteva 1971; DeGraaf and Wentworth 1986; Thompson 1993; Jokimäki et al. 1996; Jokimäki and Suhonen 1998; Fernández-Juricic and Jokimäki 2001; Chamberlain et al. 2007; Jim and Chen 2008, 2009; Ortega-Álvarez and MacGregor-Fors 2009; Caula et al. 2010, 2013).

5.3 Role of 'Gray' and 'Green' Infrastructure on Birds in Latin America

In this section, we review a substantial sample of the literature focused on the documented relationships between both 'gray' and 'green' infrastructures and birds in urban Latin America. For this, we performed an intensive search for published studies in Web of Science platform (www.webofknowledge.com), several other databases (i.e., Google Scholar, Scopus, Science Direct, PubMed), as well as regional journals and theses, using the following keywords: 'urban', 'bird', and 'ecology' in both Spanish and English. Additionally, we searched for publications in the references of the retrieved publications, especially literature reviews. As a result of our search, we gathered a total of 90 studies conducted in 14 Latin American countries.

It is noteworthy that both 'gray' and 'green' infrastructures in urban Latin America are the products of a transcultural process highly influenced by Western colonization, excluding preexisting natural landscapes and focused on European traditions (Faggi and Ignatieva 2009). Currently, Latin American cities are compact, with high population densities and highly unequal socioeconomically and environmentally (Ingram and Carroll 1981; Psacharopoulos et al. 1995; Pauchard and Barbosa 2013; Escobedo et al. 2015). The latter shows the complexity of the context in which urban ecology is done in the region, with both 'gray' and 'green' infrastructures often separated from each other, with the latter mostly focused on human use, rather than to safeguard ecological processes or provide habitat for wildlife species.

5.3.1 *'Gray' Infrastructure*

Despite the fact that an important proportion of studies focused on the drivers of avian assemblage in urban Latin America has studied vegetation traits, few have addressed the 'gray' components of cities. Some of them have focused on bird communities and others on particular species that are, in general, positively related to 'gray' components. Regarding the community level, several studies have found relationships between urbanization intensity and bird species richness and abundance (see Chap. 3 for a thorough review and analysis), with highly urbanized sites sheltering less bird species but a higher number of individuals of a few urban-dwelling species (also known as 'urban exploiters'; see Blair 1996 and Fischer et al. 2015; Fig. 5.1). Some of these studies have related buildings (and their traits; e.g., height), as well as built cover, as some of the predictor variables of such decreases (Leveau and Leveau 2004; Rivera-Gutierrez 2006; Torres Pinheiro et al. 2008; Villegas and Garitano-Zavala 2010; MacGregor-Fors and Schondube 2011; Reis et al. 2012; Toledo et al. 2012; Sanz and Caula 2015). Another topic that has not received enough attention is the role of specific variables related to the urban 'gray' infrastructure. One example of such types of studies is that of MacGregor-Fors and Schondube (2011) evaluated the relationship between bird richness and abundance with 33 variables in Morelia (Mexico), of which 14 described the 'gray' infrastructure (i.e., built cover, number of electric and telephone poles, cables, lamp poles, niches, balconies, lightning rods, antennas, exposed construction rods, maximum and minimum building heights, number of windows, number of street doors). Results of this study show that highly abundant bird species (i.e., House Sparrow—*Passer domesticus*, Barn Swallow—*Hirundo rustica*, Rock Pigeon—*Columba livia*, Inca Dove—*Columbina inca*, Great-tailed Grackle—*Quiscalus mexicanus*) were positively related with cables and lightning rods, while their richness was positively

Fig. 5.1 Rock Pigeon (*Columba livia*), one of the most representative urban-related species across the globe (Photo: Ian MacGregor-Fors)

associated with building height. On the other hand, the number of moderately abundant species was negatively related with built cover, and number of electric and telephone poles. Finally, rare bird species richness was positively linked to number of electric and telephone poles, while their abundances were negatively related to the number of lamp poles. This does not only show the complexity of the relationships between bird assemblage and the 'gray' infrastructure of cities, but also shows that several specific variables are of importance for birds.

In relation to the few urban 'dwelling', often exploiter, species that are mostly generalist and/or exotic, studies have shown that their numbers increase importantly in urban scenarios, both in large urban centers (e.g., Mexico City, Buenos Aires, Bogotá, São Paulo, Rio de Janeiro, Valencia–Chile) and smaller ones (Morelia, Mar del Plata, Pelotas, Taubaté) (Ruszczyk et al. 1987; Leveau and Leveau 2004; Faggi and Perepelezin 2006; Ortega-Álvarez and MacGregor-Fors 2009; Perepelizin and Faggi 2009; MacGregor-Fors et al. 2010; Fontana et al. 2011; Toledo et al. 2012; Sacco et al. 2013; Fig. 5.2). Two widely distributed examples of the latter are the Rock Pigeon and House Sparrow (Aronson et al. 2014; Figs. 5.1 and 5.3). These species, native to Eurasia and northern Africa, are both frequent and often hyperabundant in many Latin American cities (e.g., Carbó-Ramírez and Zuria 2011; Fontana et al. 2011; Leveau 2013). Novel sites, such as derelict and abandoned buildings, are used by these species and other generalist urban 'dwellers' for perch-

Fig. 5.2 Food provisioning to Rock Pigeons (*Columba livia*) by people in Heredia (Costa Rica) (Photo: Ian MacGregor-Fors)

Fig. 5.3 House Sparrow (*Passer domesticus*) perched on a house fence in Ajijic, Jalisco (Mexico) (Photo: Juan F. Escobar-Ibáñez)

ing, roosting, and even nesting (MacGregor-Fors et al. 2010; Perepelizin et al. 2014; Villalba-Sánchez et al. 2015). Their presence and numbers are often used as indicators of highly urbanized and busy areas (de Vasconcelos 1999; Amâncio et al. 2008; Perepelizin and Faggi 2009; Hernández et al. 2010; MacGregor-Fors and Schondube 2011; Londoño-Betancourth 2013). Additionally, there is evidence that some of these species can have detrimental ecological effects, as well as human health issues, with the House Sparrow being related to declines in native bird species richness (MacGregor-Fors et al. 2010; Ortega-Álvarez and MacGregor-Fors 2010) and Rock Pigeons being related with diseases and vectors (de Sousa et al. 2010; Ortiz et al. 2015; Pérez-García et al. 2015).

5.3.2 'Green' Infrastructure

Although many urban greenspaces are mostly determined by human decisions, their presence and traits have been shown to drive the avian species that they can shelter (Leveau and Leveau 2004; Faggi and Perepelizin 2006; Caula et al. 2010; Sanz and Caula 2015). In agreement with studies from other regions across the globe, results show that Latin American urban greenspaces shelter a higher number of bird species when compared with more urbanized sites. Yet, the mere existence of urban greenspaces does not necessarily ensure potential habitats for birds with regional conditions often playing a crucial role. Below, we summarize the results of studies focused on the relationship between greenspace traits (i.e., location, size, heterogeneity, vegetation characteristics, management, and disturbances) and bird ecology in urban Latin America.

5.3.2.1 Greenspace Location

The location of greenspaces within Latin American cities has been identified as one of the main drivers of bird community diversity (Galina and Gimenes 2006). Although the effect of location of greenspaces greatly depends on the size of cities, the semipermeability process that occurs in peri-urban areas, representing an important ecological interaction zone that differs greatly from 'intra-urban' areas, limits the presence of some species that are able to dwell under given urban conditions, including greenspaces (Garaffa et al. 2009; MacGregor-Fors 2010; MacGregor-Fors and Ortega-Álvarez 2011; Puga-Caballero et al. 2014).

In southwestern Mexico City, MacGregor-Fors and Ortega-Álvarez (2011) showed that bird species richness decreased with increasing distance from a preserve located in the boundary of the city in greenspaces of similar size. Contrarily, bird densities increased, with a dramatic peak in the farthest greenspace from the border of the city. Also, taxonomic similarity decreased almost linearly in relation to the assemblage from the reference preserve. Accordingly, Urquiza and Mella (2002) recorded a higher number of bird species in greenspaces of Santiago (Chile), as the distance to the premountain Andes decreased.

A recent study performed in Xalapa, a small-sized Mexican city, showed that when comparing greenspaces of different sizes and locations within the same urban continuum, bird species richness was highest in the largest and outermost greenspace, followed by an array of medium-sized to small-sized greenspaces with different disturbance rates, and least richness in a small highly disturbed greenspace (MacGregor-Fors et al. 2016). Regarding species composition, this study shows that vegetation structure plays a more important role than size does (Chap. 3). The latter agrees with the findings of previous studies that have addressed the value of small urban greenspaces for birds. For example, Carbó-Ramírez and Zuria (2011), based on their study performed in Pachuca, a medium-sized city located in a Mexican arid-scrub region, highlight the importance of small greenspaces and the need to expand their vegetation cover to favor both resident and migrant bird species. Also, studies from Brazil and Costa Rica (Stiles 1990; Mendonça-Lima and Fontana 2000; Guix 2007) have suggested that small-to-medium-sized greenspaces, such as neighborhood parks and private gardens, can play an important role as 'habitat islands' for birds to breed or use as stopover sites during migration.

5.3.2.2 Greenspace Size and Heterogeneity

As occurs worldwide, studies focused on the relationship between urban birds and greenspace size are common in Latin America (see Chace and Walsh 2006; Evans et al. 2009; MacGregor-Fors et al. 2009; Marzluff 2016 and references therein). In general, results of these studies reinforce the existence of positive species–area relationships, suggesting that larger areas have conditions that can support more avian species, as described in the island biogeography theory (MacArthur and Wilson 1967; Urquiza and Mella 2002; Garitano-Zavala and Gismondi 2003; Faggi and

Perepelizin 2006; Carbó-Ramírez and Zuria 2011). Yet, large greenspaces with higher bird diversity are often related with heterogeneous environmental conditions, involving plant diversity and physiognomic complexity, low human disturbance rates, and management regimes, among others (MacGregor-Fors et al. 2016). However, some studies performed in Latin America have not found positive species–area relationships in urban greenspaces. For example, Malagamba-Rubio et al. (2013) did not find statistical differences in bird species richness between small (0.27–3.85 ha), medium (8.63–12 ha), and large (22.04 ha) greenspaces of Querétaro, a semi-arid Mexican city. This suggests that greenspace size per se does not necessarily determine the number of bird species that can dwell within a given urban greenspace. Due to the unfavorable conditions that highly urbanized sites often pose to many bird species, small parks, squares, private gardens, median strips, and even isolated street trees may play crucial roles for birds able to use their resources and survive their hazards (Pineda-López et al. 2010; Carbó-Ramírez and Zuria 2011); however, these spaces do not necessarily ensure the maintenance of viable populations (González-Oreja et al. 2007).

Regarding greenspace heterogeneity, several studies have shown that, regardless of bird species richness, species composition can differ in contrasting conditions of vegetation heterogeneity (Manhães and Loures-Ribeiro 2005; Gómez 2006; Lopes and Anjos 2006; Caula et al. 2010). The latter does not only apply for single greenspaces, but also for urban greenspace networks, where the contribution to the total diversity increases when the array of greenspaces is diverse in terms of size, heterogeneity, vegetation traits, management, and human disturbances (Faggi and Perepelizin 2006; Cursach and Rau 2008; MacGregor-Fors et al. 2016). Caula et al. (2010), in Valencia City (Venezuela), compared two close areas of the same size embedded within the same urban matrix, but differing in complexity: the first area was highly heterogeneous, while the second one was simplified by human management. Results of this study show that bird richness was two times higher in the first area, sheltering species considered as urban 'avoiders' (e.g., Red Rumped Woodpecker—*Veniliornis kirkii*, Lineated Woodpecker—*Dryocopus lineatus*, Boat-billed Heron—*Cochlearius cochlearius*, Green Kingfisher—*Chloroceryle americana*). This study highlights the importance of native remnant areas that preserved vegetation heterogeneity for urban bird conservation. Similar findings have been reported comparing municipal greenspaces maintained for recreational purposes with unmaintained greenspaces in the city of Valdivia, in Southern Chile (Silva et al. 2015).

5.3.2.3 Greenspace Vegetation Traits

Greenspaces in urban Latin America are highly diverse in origin, type, and traits. A first classification of such diversity has to do with their use. Commonly, urban greenspaces are categorized as conservation remnants, parks, university campuses, sports fields, gardens, median strips, among others. Although many urban bird ecological studies have been carried out in urban parks, the amount of studies developed in university campuses is noteworthy (e.g., Escobar and Salomón 1983; MacGregor-Fors 2008; Pozo and Cisneros 2014). Although greenspace type can be a proxy of human use and management, studies have focused on addressing the role that specific greenspace variables have on urban bird communities.

Vegetation traits have been identified as the local-scale greenspace variables that drive urban bird communities. Among plants, trees have also shown to play crucial roles for bird diversity in greenspaces, and among tree traits, their cover, diversity, and size (measured as height and/or diameter at breast height) have shown to be positively related with bird diversity (e.g., Naranjo and Estela 1999; Garitano-Zavala and Gismondi 2003; MacGregor-Fors 2008). Such local-scale vegetation variables have often shown to override the importance of greenspace size and location—mainly isolation from the city border—in urban Latin America (Urquiza and Mella 2002; Toledo et al. 2012). Yet, there is an important bias toward urban tree planting, rather than including native vegetation components to their greenspaces. A study performed by Leveau and Leveau (2004) shows that regions whose original systems lack trees (e.g., Argentinean Pampas) are mainly comprised by shrub and grass-related birds. Including urban greenspaces that differ from the original ecosystems, such as tree-based ones in arid regions, can attract different species to those present in the immediate species pool. The latter was recorded in the city of Pachuca, where eight species that had not been recorded in several years in arid tropical scrubs in and around the city were recorded in tree-based greenspaces (Carbó-Ramírez and Zuria 2011).

Another vegetation trait that has been controversial in urban Latin America is the role of species origin (namely native and exotic) on bird diversity (Fig. 5.4). Particularly for exotic vegetation, studies have shown both positive and negative relationships with bird diversity. For example, Noguera Chacón (2012) compared two greenspaces differing in size and vegetation origin in Quito (Ecuador), finding that the smaller one with native vegetation remnants had more than twice bird species richness than the larger one, which was mainly covered by gum and pine trees. In contrast, a larger number of birds were recorded in mixed native–exotic woodland of an heterogeneous greenspace of Mendoza (Argentina) when compared with a native scrubland (Gómez 2006). Other studies have shown scenarios under which exotic plants can attract bird species, most of which have been related with food provisioning. For instance, some exotic trees in Argentina (e.g., *Fraxinus* spp., *Liquidambar styraciflua, Ulmus procera, Ligustrum lucidum*) produce abundant seeds that attract native birds to cities. This is the case of the Green Ash (*Fraxinus pennsylvanica*), which has been related with the arrival and fairly commonness of the Hooded Siskin (*Spinus magellanicus*) in Buenos Aires, Argentina (Montaldo and Roitman 2000).

Exotic trees with nectar-rich flowers, such as Australian Silk Oaks (*Grevillea robusta*) in Guadalajara (Mexico), have also been found to attract both breeding and wintering bird species (MacGregor-Fors 2008). Fruit trees have also been recorded to attract birds in urban Latin America, which are widespread through cities and its greenspaces and provide important ecoservices (Dobbs et al. in rev). A study focused on fruit-eating birds in 11 large and medium-sized Brazilian cities found that many of the migratory species feed on exotic urban trees, mainly during winter, while others suggest that exotic trees represent year-round food resources for birds

Fig. 5.4 Greenspace located in peri-urban Bogotá (Colombia) with a sign promoting the protection of native vegetation (Photo: Juan F. Escobar-Ibáñez)

(Santos and Ragusa-Netto 2013; Oliveira et al. 2013). Although this could be considered as positive for birds, it is tricky (as generally occurs with exotic, potentially invasive, and invasive plant species; Vitousek et al. 1997; Corlett 2005). Birds could be important dispersers of nonnative plants in other human-modified systems, as well as native ecosystems, posing a potential ecological threat at different scales (Guix 1996, 2007; Montaldo and Roitman 2000). Also, a recently published study from São Carlos (Brazil) showed exotic tree abundance to be positively related with bird richness in its university campus, but note that native tree species were clearly underrepresented in the area (Lessi et al. 2016).

5.3.2.4 Greenspace Management and Human Activities

The way in which urban spaces are designed and managed can influence the birds that can inhabit them (Evans et al. 2009). In particular, vegetation management through mowing and pruning, among other human activities (e.g., path clearing, leaf sweeping, passing pedestrians, picnicking), can play crucial roles on the bird diversity that urban greenspaces can shelter. Vegetation management practices regularly alter vegetation physiognomy, often decreasing its structural complexity and limiting the presence of specialist bird species (Leveau and Leveau 2004; Filloy et al. 2015). As noted earlier, Caula et al. (2003) compared two similar-sized greenspaces: a native vegetation remnant and a managed park, finding that bird species richness was twofold in the native remnant with many urban avoider species. However, open and disturbed areas in greenspaces do create novel areas of which some bird species take advantage. As outlined in two regional reviews that include Latin American cities, open-space insectivores and granivores have been highly successful in urban Latin America, often representing the predominant groups within its cities (Ortega-Álvarez and MacGregor-Fors 2011a, b).

Regarding human activities, several studies have negatively related recreational activities and disturbances, caused by pedestrians and vehicles, with avian diversity in Latin American cities. In the city of Valdivia (Chile), Silva et al. (2015) found that greenspaces with recreational purposes shelter less bird diversity than nonmanaged ones. Also, studies performed in two Mexican cities (i.e., Morelia, Mexico City) show negative relationships between passing cars and bird species richness (Ortega-Álvarez and MacGregor-Fors 2009; MacGregor-Fors et al. 2010, 2011). Moreover, passing pedestrians were identified as a predictor variable related to decreases in the abundance of several bird species in greenspaces of Mexico City (e.g., House Finch–*–Haemorhous mexicanus*, Rufous-backed Robin—*Turdus rufopalliatus*). Leveau (2008) assessed the spatial and temporal variations of nectar feeding in Krantz Aloe (*Aloe arborescens*) by House Sparrows in a public garden and a recreational area along the coast of Mar del Plata (Argentina), finding that vehicle and pedestrian traffic were negatively related with sparrow visitation rates during weekends.

5.4 Opportunities and Future Research Needs

Research developed in urban Latin America has shown that birds respond to the urbanization process both similarly and differently to the patterns reported for other regions. Due to its high biodiversity and urbanization rates, Latin America is a clear target for conservation, for which evidence-based information of the effect of urbanization on birds is crucial (Ortega-Álvarez and MacGregor-Fors 2011a, b). One main gap regarding urban infrastructure, basically given by a greenspace survey bias, has to do with our knowledge of the ecological patterns and processes occurring in highly developed urbanized systems. In general, studies lack comparisons with the 'urban matrix', failing to set a baseline of its effect assessed through paired comparisons (see MacGregor-Fors and Schondube 2011; MacGregor-Fors et al. 2015, 2016). Such contrasts could shed important light on many ecological topics, ranging from community diversity drivers, to ethological, physiological, health, and even genetic assessments of the ways in which birds respond to urbanization, allowing to make the leap toward more mechanistic urban ecology studies (as suggested by Shochat et al. 2006 and McDonnell and Hahs 2013).

Also regarding the 'gray' infrastructure, there is an urgent need to understand the role of socioeconomics, as well as the types and intensities of 'gray' urban infrastructure (e.g., industrial, residential, commercial), with all its physical, ecological, and social components, not only as comparison standpoints for greenspaces, but also as the constant matrix of urban landscapes. Studies from across the globe, including Latin America (e.g., MacGregor-Fors and Shondube 2011; Reis et al. 2012; Chávez-Zichinelli et al. 2013; Silva et al. 2015), have shown that these variables may play imperative roles in determining the response of birds in urban systems. There are other variables of the 'gray' infrastructure that represent important threats to bird diversity, such as window strikes (see Chap. 7 for a thorough review on the topic); yet, our understanding of their effects in the region is woefully limited.

Although urban 'green' infrastructure has been widely studied in Latin America, there are important areas within the region for which we remain to ignore basic ecological information of cities. Thus, we urge ecologists to study their cities, including both 'gray' and 'green' infrastructures in the best of cases, and moving forward with the rest of well-studied regions to improve our understanding on their role as avian drivers in urban Latin America. Long-term studies considering the wide array of 'greenspaces' (including some widely ignored ones, such as abandoned lots) will allow the generated knowledge translate to action. One urgent topic regarding 'green' infrastructure has to do with the origin of the used plants within and across streetscapes. Although in some cases exotic plants have been shown to play a positive role for birds, there is evidence of contrasting results that have highlighted the potential threats of using them to populate urban areas. Thus, a major discussion must be established between ecologists, botanists, landscape architects, and local governments, to say the least, to tackle this highly complex issue.

Undoubtedly, studies have focused on the most conspicuous infrastructure of cities, namely the 'gray' and the 'green' ones, failing to include other conditions that have been widely ignored. For instance, water bodies ('blue' infrastructure) have been shown to be highly relevant bird diversity in urban Latin America (e.g., Bryce and Hughes 2002; Lugo 2002; Rosselli and Stiles 2012a, b; Pineda-López 2011; Lobo et al. 2015). Furthermore, there are other types of urban conditions that need to be cautiously analyzed if we aim to understand cities as systems, such as wastelands (e.g., deposits, treatment ponds). Due to their physical, ecological, and social relevance, as well as the uniqueness of their role in food provisioning and relationship with urban 'dweller' wildlife species, these sites, which we suggest to name 'brown' infrastructure, are interesting to be assessed using birds as bioindicators at many scales (Yorio and Giaccardi 2002; Novaes and Cintra 2015; Ortiz et al. 2015).

References

Alberti M (2005) The effects of urban patterns on ecosystem function. Int Reg Sci Rev 28:168–192

Amâncio S, Souza VB, Melo C (2008) *Columbia livia* e *Pitangus sulphuratus* como indicadoras de qualidade ambiental em área urbana. Rev Bras Ornitol 16:32–37

Andersson E, Barthel S, Ahrné K (2007) Measuring social-ecological dynamics behind the generation of ecosystem services. Ecol Appl 17:1267–1278

Aronson MFJ, La Sorte FA, Nilon CH et al (2014) A global analysis of the impacts of urbanization on bird and plant diversity reveals key anthropogenic drivers. Proc R Soc Lond B Biol Sci 281:1–8

Avolio ML, Pataki DE, Pincetl S et al (2015) Understanding preferences for tree attributes: the relative effects of socio-economic and local environmental factors. Urban Ecosyst 18:73–86

Blair RB (1996) Land use and avian species diversity along an urban gradient. Ecol Appl 6:506–519

Bryce SA, Hughes RM (2002) Variable assemblage responses to multiple disturbance gradients: Oregon and Appalachia, USA, case studies. In: Simon TP (ed) Biological response signatures: multimetric index patterns for assessment of freshwater assemblages. CRC Press, Boca Raton, pp 539–560

Carbó-Ramírez P, Zuria I (2011) The value of small urban greenspaces for birds in a Mexican city. Landsc Urban Plan 100:213–222

Caula S, De Nóbrega JR, Giner S (2003) La diversidad de aves como elemento de una estrategia de conservación del Jardín Botánico de Valencia, Venezuela. Acta Biol Venez 23:1–13

Caula S, Giner S, De Nóbrega JR (2010) Aves urbanas: Un estudio comparativo en dos parques tropicales con diferente grado de intervención humana (Valencia, Venezuela). Faraute 5:2

Caula S, Villalobos A, Marty P (2013) Seasonal dynamics of bird communities in urban forest of a Mediterranean city (Montpellier, Southern France). Urban Ecosyst 17:11–26

Chace JF, Walsh JJ (2006) Urban effects on native avifauna: a review. Landsc Urban Plan 74:46–69

Chamberlain DE, Cannon AR, Toms MP (2004) Associations of garden birds with gradients in garden habitat and local habitat. Ecography 27:589–600

Chamberlain DE, Gough S, Vaughan H et al (2007) Determinants of bird species richness in public green spaces. Bird Study 54:87–97

Chávez-Zichinelli CA, MacGregor-Fors I, Quesada J et al (2013) How stressed are birds in an urbanizing landscape? Relationships between the physiology of birds and three levels of habitat alteration. Condor 115:84–92

Corlett RT (2005) Interactions between birds, fruit bats and exotic plants in urban Hong Kong, South China. Urban Ecosyst 8:275–283

Croci S, Butet A, Clergeau P (2008) Does urbanization filter birds on the basis of their biological traits? Condor 110:223–240

Cursach J, Rau J (2008) Avifauna presente en dos parques urbanos de la ciudad de Osorno, Sur de Chile. Bol Chil Ornitol 14:98–103

Davies AM, Glick TF (1978) Urban ecosystems and island biogeography. Environ Conserv 5:299–304

de Sousa E, Berchieri Júnior A, Pinto AA et al (2010) Prevalence of *Salmonella* spp. antibodies to *Toxoplasma gondii*, and Newcastle disease virus in feral pigeons (*Columba livia*) in the city of Jaboticabal, Brazil. J Zoo Wildlife Med 41:603–607

de Vasconcelos MF (1999) Contribuição ao conhecimento ornitológico do Pico do Papagaio, município de Aiuruoca, Minas Gerais. Atual Ornitol 90:10–11

DeGraaf RM, Wentworth JM (1986) Avian guild structure and habitat associations in suburban bird communities. Urban Ecol 9:399–412

Donovan GH, Butry DT (2010) Trees in the city: valuing street trees in Portland, Oregon. Landsc Urban Plan 94:77–83

Eldredge N, Horenstein S (2014) Concrete jungle: New York city and our last best hope for a sustainable future. University of California Press, Oakland

Emlen JT (1974) An urban bird community in Tucson, Arizona: derivation, structure, regulation. Condor 76:184–197

Escobar-Ibáñez JF, MacGregor-Fors I (2015) On the tightrope: use of open sky urban telephone wires by Azure-crowned Hummingbirds (*Amazilia cyanocephala*) for nesting. Wilson J Ornithol 127:297–302

Escobar ME, Salomón MJ (1983) Contribución al conocimiento de algunas especies de aves que habitan el Campus Universitario de San Lorenzo-Paraguay. I Symposio de Ornitología Neotropical Arequipa, Perú, 14–15 Oct 1983
Escobedo FJ, Nowak D, Wagner J et al (2006) The socioeconomics and management of Santiago de Chile's public urban forests. Urban For Urban Gree 4:105–114
Escobedo FJ, Clerici N, Staudhammer CL et al (2015) Socio-ecological dynamics and inequality in Bogotá, Colombia's public urban forests and their ecosystem services. Urban For Urban Gree 14:1040–1053
Evans KL, Gaston KJ, Frantz AC et al (2009) Independent colonization of multiple urban centers by a formerly forest specialist bird species. Proc R Soc Lond B Biol Sci 276:2403–2410
Faggi A, Ignatieva M (2009) Urban green spaces in Buenos Aires and Christchurch. Munic Eng 162:241–250
Faggi A, Perepelizin P (2006) Riqueza de aves a lo largo de un gradiente de urbanización en la ciudad de Buenos Aires. Mus Argentino Cienc Nat ns 8:289–297
Fernández-Juricic E, Jokimäki J (2001) A habitat island approach to conserving birds in urban landscapes: case studies from southern and northern Europe. Biodivers Conserv 10:2023–2043
Filloy J, Grosso S, Bellocq MI (2015) Urbanization altered latitudinal patterns of bird diversity – environment relationships in the southern Neotropics. Urban Ecosyst 18:777–791
Fischer JD, Schneider SC, Ahlers AA et al (2015) Categorizing wildlife responses to urbanization and conservation implications of terminology. Conserv Biol 29:1246–1250
Fontana CS, Burger MI, Magnusson WE (2011) Bird diversity in a subtropical South-American City: effects of noise levels, arborisation and human population density. Urban Ecosyst 14:341–360
Forrest M, Konijnendijk C (2005) A history of urban forests and trees in Europe. In: Konijnendijk CC, Nilsson K, Randrup TB et al (eds) Urban forests and trees. Springer, Berlin, pp 23–48
Galina AB, Gimenes MR (2006) Riqueza, composição e distribuição espacial da comunidade de aves em um fragmento florestal urbano em Maringá, Norte do Estado do Paraná, Brasil. Acta Scientiarum 28:379–388
Garaffa PI, Filloy J, Bellocq MI (2009) Bird community responses along urban–rural gradients: does the size of the urbanized area matter? Landsc Urban Plan 90:33–41
Garitano-Zavala A, Gismondi P (2003) Variación de la riqueza y diversidad de la ornitofauna en áreas verdes urbanas de las ciudades de La Paz y El Alto (Bolivia). Ecología en Bolivia 38:65–78
Gómez-Baggethun E, Barton DN (2013) Classifying and valuing ecosystem services for urban planning. Ecol Econ 86:235–245
Gómez VE (2006) Aves del Parque General San Martín (Mendoza). Distribución y características Multequina 15:81–95
González-Oreja JA, Bonache C, Buzo D et al (2007) Caracterización ecológica de la avifauna de los parques urbanos de la ciudad de Puebla (México). Ardeola 54:53–67
Guix JC (1996) Áspectos da frugivoria, disseminação e predação de sementes por vertebrados nas florestas nativas do estado de São Paulo, sudeste do Brasil. PhD dissertation, Universitat de Barcelona
Guix JC (2007) The role of alien plants in the composition of fruit-eating bird assemblages in Brazilian urban ecosystems. Orsis 22:87–104
Hadidian J, Sauer J, Swarth C et al (1997) A citywide breeding bird survey for Washington, DC. Urban Ecosyst 1:87–102
Hernández A, Estela FA, Chacón de Ulloa P (2010) ¿Es *Sayornis nigricans* (Aves: Tyrannidae) un buen indicador de calidad ambiental en la zona urbana de Cali, Colombia? Rev Acad Colomb Cienc 34:373–380
Ingram GK, Carroll A (1981) The spatial structure of Latin American cities. J Urban Econ 9:257–273
Jim CY, Chen WY (2008) Assessing the ecosystem service of air pollutant removal by urban trees in Guangzhou (China). J Environ Manag 88:665–676

Jim CY, Chen WY (2009) Ecosystem services and valuation of urban forests in China. Cities 26:187–194
Jokimäki J, Suhonen J (1998) Distribution and habitat selection of wintering birds in urban environments. Landsc Urban Plan 39:253–263
Jokimäki J, Suhonen J, Inki K et al (1996) Biogeographical comparison of winter bird assemblages in urban environments in Finland. J Biogeogr 23:379–386
Konijnendijk CC (2016) Tree agency and urban forest governance. SASBE 5:176–188
Lessi BF, Rodrigues-Pires JS, Batisteli AF et al (2016) Vegetation, urbanization, and bird richness in a Brazilian periurban area. Ornitol Neotrop 27:203–210
Leveau CM, Leveau LM (2004) Comunidades de aves en un gradiente urbano de la ciudad de Mar del Plata Argentina. El Hornero (B. Aires) 19:13–21
Leveau L (2008) Dynamics of nectarivory in the House Sparrow in an urban environment. Ornitol Neotrop 19:275–281
Leveau LM (2013) Bird traits in urban–rural gradients: how many functional groups are there? J Ornithol 154:655–662
Lobo N, Faggi A, de Cabo L (2015) Indicadores ecológicos en el monitoreo ambiental de lagunas restauradas: El caso Costanera Sur. Ingenium UNLZ 3:6–11
Londoño-Betancourth JC (2013) Discusiones sobre la presencia de aves rapaces, aves migratorias y aves bajo algún grado de amenaza en la ciudad de Pereira, Risaralda. Rev Luna Azul 36:134–164
Lopes EV, Anjos L (2006) A composição da avifauna do campus da Universidade Estadual de Londrina, norte do Paraná, Brasil. Rev Bras Zool 23:145–156
Lugo AE (2002) Conserving Latin American and Caribbean mangroves: issues and challenges. Madera Bosques 1:5–25
MacArthur RH, Wilson EO (1967) The theory of island biogeography. Princeton University Press, Princeton
MacGregor-Fors I (2008) Relation between habitat attributes and bird richness in a western Mexico suburb. Landsc Urban Plan 84:92–98
MacGregor-Fors I (2010) How to measure the urban-wildland ecotone: redefining "peri-urban" areas. Ecol Res 25:883–887
MacGregor-Fors I (2011) Misconceptions or misunderstandings? On the standardization of basic terms and definitions in urban ecology. Landsc Urban Plan 100:347–349
MacGregor-Fors I, Ortega-Álvarez R (2011) Fading from the forest: bird community shifts related to urban park site-specific and landscape traits. Urban For Urban Gree 10:239–246
MacGregor-Fors I, Schondube JE (2011) Gray vs. green urbanization: relative importance of urban features for urban bird communities. Basic Appl Ecol 12:372–381
MacGregor-Fors I, Ortega-Álvarez R, Schondube JE (2009) On the ecological quality of urban systems: an ornithological perspective. In: Graber DS, Birmingham KA (eds) Urban planning in the 21st century. Nova Science Publishing, NewYork, pp 51–66
MacGregor-Fors I, Morales-Pérez L, Quesada J et al (2010) Relationship between the presence of House Sparrows (*Passer domesticus*) and Neotropical bird community structure and diversity. Biol Invasions 12:87–96
MacGregor-Fors I, Morales-Pérez L, Schondube JE (2011) Does size really matter? Species–area relationships in human settlements. Divers Distrib 17:112–121
MacGregor-Fors I, Morales-Pérez L, Schondube J (2012) From forests to cities: effects of urbanization on tropical birds. In: Lepczyk CA, Warren PS (eds) Urban bird ecology and conservation. University of California Press, Berkeley, pp 33–48
MacGregor-Fors I, Avendaño-Reyes S, Bandala VM et al (2015) Multi-taxonomic diversity patterns in a neotropical green city: a rapid biological assessment. Urban Ecosyst 18:633–647
MacGregor-Fors I, Escobar F, Rueda-Hernández R et al (2016) City "green" contributions: the role of urban greenspaces as reservoirs for biodiversity. Forests 7:146
Malagamba-Rubio A, MacGregor-Fors I, Pineda-López R (2013) Comunidades de aves en áreas verdes de la ciudad de Santiago de Querétaro, México. Ornitol Neotrop 24:371–386

Manhães MA, Loures-Ribeiro A (2005) Spatial distribution and diversity of bird community in urban area of southeast Brazil. Braz Arch Biol Technol 48:285–294

Marzluff JM (2016) A decadal review of urban ornithology and a prospectus for the future. Ibis 159:1–13

McDonnell MJ, Hahs AK (2008) The use of gradients analysis studies in advancing our understanding of the ecology of urbanizing landscapes: current status and future directions. Landsc Ecol 23:1143–1155

McDonnell MJ, Hahs AK (2013) The future of urban biodiversity research: moving beyond the 'low-hanging fruit'. Urban Ecosyst 16:397–409

McDonnell MJ, Hahs AK (2015) Adaptation and adaptedness of organisms to urban environments. Annu Rev Ecol Evol S 46:261–280

McDonnell MJ, Pickett ST, Groffmam P et al (1997) Ecosystem processes along an urban-to-rural gradient. Urban Ecosyst 1:21–36

McKinney ML (2008) Effects of urbanization on species richness: a review of plants and animals. Urban Ecosyst 11:161–176

Mendonça-Lima A, Fontana CS (2000) Composição, freqüência e aspectos biológicos da avifauna no Porto Alegre Country Clube, Rio Grande do Sul. Ararajuba 8:1–8

Montaldo NH, Roitman GG (2000) Plantas no-asteraceae en la alimentación del cabecita negro común (*Carduelis magellanica*). El Hornero (B Aires) 15:99–102

Naranjo LG, Estela F (1999) Inventario de la avifauna de un área suburbana de la ciudad de Cali. Boletín SAO X:11–27

Noguera Chacón T (2012) Evaluación preliminar del estado de conservación del Parque Metropolitano y del Parque Itchimbía. BSc Thesis, Universidad San Francisco de Quito, Ecuador

Novaes WG, Cintra R (2015) Anthropogenic features influencing occurrence of Black Vultures (*Coragyps atratus*) and Turkey Vultures (*Cathartes aura*) in an urban area in central Amazonian Brazil. Condor 117:650–659

Nowak DJ, Greenfield EJ (2012) Tree and impervious cover in the United States. Landsc Urban Plan 107:21–30

Nuorteva P (1971) The synanthropy of birds as an expression of the ecological cycle disorder caused by urbanization. Ann Zool Fenn 8:547–553

Oliveira DSF, Franchin AG, Marçal-Júnior O (2013) Disponibilidade de frutos de *Michelia champaca* L. (Magnoliaceae) e seu consumo por aves na área urbana de Uberlândia, MG. Biosci J 29:2053–2065

Ortega-Álvarez R, MacGregor-Fors I (2009) Living in the big city: effects of urban land-use on bird community structure, diversity, and composition. Landsc Urban Plan 90:189–195

Ortega-Álvarez R, MacGregor-Fors I (2010) What matters most? Relative effect of urban habitat traits and hazards on urban park birds. Ornitol Neotrop 21:519–533

Ortega-Álvarez R, MacGregor-Fors I (2011a) Spreading the word: the ecology of urban birds outside the United States, Canada, and Western Europe. Auk 128:415–418

Ortega-Álvarez R, MacGregor-Fors I (2011b) Dusting-off the file: a review of knowledge on urban ornithology in Latin America. Landsc Urban Plan 101:1–10

Ortiz SG, Pastrán MG, Blanco Fager RV et al (2015) Impacto provocado por dormideros de aves en Ciudad Capital de San Juan-Argentina. Caso Peatonal de Microcentro. In: Actas VII Jornadas Regionales y 5° Nacionales de ecología urbana-verde urbano, La Rioja, 14–16 September 2015

Pauchard A, Barbosa O (2013) Regional assessment of Latin America: rapid urban development and social economic inequity threaten biodiversity hotspots. In: Elmqvist T, Fragkias M, Goodness J et al (eds) Urbanization, biodiversity and ecosystem services: challenges and opportunities. A global assessment. Springer, Dordrecht, pp 589–608

Pedlowski M, Corabi J, Heynen N (2002) Urban forest and environmental inequality in Campos dos Goytacazes, Rio de Janeiro, Brazil. Urban Ecosyst 6:9–20

Pellissier V, Cohen M, Boulay A et al (2012) Birds are also sensitive to landscape composition and configuration within the city centre. Landsc Urban Plan 104:181–188

Perepelezin P, Faggi A (2009) Diversidad de aves en tres barrios de la ciudad de Buenos Aires. Multequina 18:71–85
Perepelezin P, Scali F, Faggi A et al (2014) Aves indicadoras del estado ambiental urbano. Gerencia Ambiental XX:16–23
Pérez-García J, Monsalve-Arcila D, Márquez-Villegas C (2015) Presencia de parásitos y enterobacterias en palomas ferales (*Columba livia*) en áreas urbanas en Envigado, Colombia. Rev Fac Nac Salud Pública 33:370–376
Pineda-López R (2011) Aves acuáticas urbanas y extra-urbanas en Querétaro, México. El Canto del Centzontle 2:87–103
Pineda-López R, Febvre N, Martínez M (2010) Importancia de proteger pequeñas áreas periurbanas por su riqueza avifaunística: El caso de Mompaní, Querétaro, México. Huitzil 11:69–80
Pozo WE, Cisneros SL (2014) Guía de aves. Biodiversidad Urbana. Aves en la ESPE. Universidad de las Fuerzas Armadas (ESPE), Quito Ecuador
Psacharopoulos G, Morley S, Fiszbein A et al (1995) Poverty and income inequality in Latin America during the 1980s. Rev Income Wealth 41:254–264
Puga-Caballero A, MacGregor-Fors I, Ortega-Álvarez R (2014) Birds at the urban fringe: avian community shifts in different peri-urban ecotones of a megacity. Ecol Res 29:619–628
Reis E, López-Iborra GM, Pinheiro RT (2012) Changes in bird species richness through different levels of urbanization: implications for biodiversity conservation and garden design in Central Brazil. Landsc Urban Plan 107:31–42
Rivera-Gutiérrez HF (2006) Composición y estructura de una comunidad de aves en un área suburbana en el suroccidente colombiano. Ornitol Colomb 4:28–38
Rosselli L, Stiles FG (2012a) Local and landscape environmental factors are important for the conservation of endangered wetland birds in a high Andean plateau. Waterbirds 35:453–469
Rosselli L, Stiles FG (2012b) Wetland habitats of the Sabana de Bogotá Andean Highland Plateau and their birds. Aquat Conserv 22:303–317
Ruszczyk A, Rodrigues JJS, Roberts TMT et al (1987) Distribution patterns of eight bird species in the urbanization gradient of Porto Alegre, Brazil. Cienc Cult 39:14–19
Sacchi R, Gentilli A, Razzetti E et al (2002) Effects of building features on density and flock distribution of feral pigeons *Columba livia* var. *domestica* in an urban environment. Can J Zool 80:48–54
Sacco AG, Bergmann FB, Rui AM (2013) Assembleia de aves na área urbana do município de Pelotas, Rio Grande do Sul, Brasil. Biota Neotrop 13:153–162
Sandström UG, Angelstam P, Mikusinski G (2006) Ecological diversity of birds in relation to the structure of urban green space. Landsc Urban Plan 77:39–53
Santos AA, Ragusa-Netto J (2013) Toco-toucan (*Ramphastos toco*) feeding habits at an urban area in Central Brazil. Ornitol Neotrop 24:1–13
Sanz V, Caula S (2015) Assessing bird assemblages along an urban gradient in a Caribbean island (Margarita, Venezuela). Urban Ecosyst 18:729–746
Seto KC, Fragkias M, Güneralp B et al (2011) A meta-analysis of global urban land expansion. PLoS One 6:e23777
Shochat E, Warren PS, Faeth SH et al (2006) From patterns to emerging processes in mechanistic urban ecology. Trends Ecol Evol 21:186–191
Silva CP, García CE, Estay SA et al (2015) Bird richness and abundance in response to urban form in a Latin American city: Valdivia, Chile as a case study. PLoS One 10:e0138120
Sodhi NS, Briffett C, Kong L et al (1999) Bird use of linear areas of a tropical city: implications for park connector design and management. Landsc Urban Plan 45:123–130
Souza FL (2016) Booking.com®: a fake mega arboreal termitaria as an unusual nesting site for the Peach-fronted parakeet, *Eupsittula aurea* (Psittacidae). Biotemas 29:69–72
Stiles G (1990) La avifauna de la Universidad de Costa Rica y sus alrededores a través de veinte años (1968-1989). Rev Biol Trop 38:361–381
Strohbach MW, Lerman SB, Warren PS (2013) Are small greening areas enhancing bird diversity? Insights from community–driven greening projects in Boston. Landsc Urban Plan 114:69–79

Thompson FR (1993) Simulated responses of a forest–interior bird population to forest management options in central hardwood forests of the United States. Conserv Biol 7:325–333

Toledo MCB, Donatelli RJ, Batista GT (2012) Relation between green spaces and bird community structure in an urban area in Southeast Brazil. Urban Ecosyst 15:111–131

Torres Pinheiro R, Dornas T, Santos Reis E et al (2008) Birds of the urban area of Palmas, TO: composition and conservation. Rev Bras Ornitol 16:339–347

Tratalos J, Fuller RA, Evans KL et al (2007) Bird densities are associated with household densities. Glob Change Biol 13:1685–1695

Urquiza A, Mella JE (2002) Riqueza y diversidad de aves en parques de Santiago durante el período estival. Bol Chil Ornitol 9:12–21

Vázques-Fuentes AE (2008) Vegetación urbana y desigualdades socioeconómicas en la comuna de Peñalolén, Santiago de Chile. Una perspectiva de justicia ambiental. MSc thesis, Universidad de Chile

Villalba-Sánchez C, De La Ossa lacayo A, De La Ossa VJ (2015) Densidad de paloma doméstica (*Columba livia domestica* GMELIN, 1789) en el antiguo mercado público de Sincelejo, Sucre, Colombia. Rev Asoc Col Cienc Biol 27:72–79

Villegas M, Garitano-Zavala A (2010) Bird community responses to different urban conditions in La Paz, Bolivia. Urban Ecosyst 13:375–391

Vitousek PM, Mooney HA, Lubchenco J et al (1997) Human domination of Earth's ecosystems. Science 277:494–499

Yorio P, Giaccardi M (2002) Urban and fishery waste tips as food sources for birds in northern coastal Patagonia, Argentina. Ornitol Neotrop 13:283–292

Chapter 6
Stay or Leave? Avian Behavioral Responses to Urbanization in Latin America

César González-Lagos and Javier Quesada

Abstract Behavioral adjustments are at the forefront of the mechanisms that birds employ to deal with environmental changes. We here review the literature focused on how behavior influences bird responses when faced with the challenges of urbanization in Latin America. Most of reviewed studies assessed for patterns of responses to urbanization with incipient information regarding the behavioral adjustments, as well as the filtering of specific behavioral phenotypes. A common and unsurprising tendency was that several avian species across Latin America are using resources from urban vegetation patches. The few experimental studies performed in urban Latin America focus on the role of personality in adjustments of foraging behavior, as well as the response to noise pollution. Nevertheless, we found no study to directly assess whether or not behavioral adjustments are related to fitness. Even so, studies assessing for the role of behavioral responses to urbanization that explicitly consider their effect on population dynamics are lacking worldwide, despite their importance for fully understanding the differential fate of species having to live in an increasingly built-up planet. Our review allowed us to identify important knowledge gaps of topics related to avian behavioral responses to urbanization, of which the following head the list: (i) behavioral adjustments in both urban greenspaces and highly developed areas; (ii) adaptiveness of avian behavioral adjustments through population dynamics; (iii) metapopulations as one of the process underlying the viability of avian bird populations; and (iv) the role of behavioral changes on evolutionary process in urban areas.

C. González-Lagos (✉)
Centro de Investigación en Recursos Naturales y Sustentabilidad (CIRENYS), Universidad Bernardo O'Higgins, Santiago, Chile

Center of Applied Ecology and Sustainability (CAPES), Pontificia Universidad Católica de Chile, Santiago, Chile
e-mail: cesar.glagos@gmail.com

J. Quesada
Department of Chordates, Natural History Museum of Barcelona, Barcelona, Catalonia, Spain

Departament Biologia Animal, Vegetal i Ecologia, Facultat de Biociències, Universitat Autònoma de Barcelona, Bellaterra, Catalonia, Spain

I. MacGregor-Fors, J.F. Escobar-Ibáñez (eds.), *Avian Ecology in Latin American Cityscapes*, DOI 10.1007/978-3-319-63475-3_6

Keywords Bird behavior • Personality • Plasticity • Risk-taking • Urban ecology

6.1 Urbanization and Bird Behavior

All animals are behaving in one way or another all the time; hence, the task of understanding the role of behavior in how animals adapt to new environments is fundamental due to its relationship with a plethora of aspects related to their ecology (e.g., migration, social behavior, reproduction, feeding, use of space and time, communication, sexual selection). Birds are often used as models for studying animal behavior and, even though the challenge of adjusting to novel environments has been widely studied, many questions still remain and prevent a full understanding of how birds confront life in urban areas and which are the related consequences (Candolin and Wong 2012; Gil and Brumm 2014; Marzluff 2016). For instance, avian behavioral adjustments when faced with a challenging situation have been shown to be essential for explaining whether or not a species is able to cope with the urbanization process (Kark et al. 2007; Lowry et al. 2013; Sol et al. 2013a). Nevertheless, some bird species that thrive in urban environments are not particularly behaviorally flexible, suggesting that there is more than one good strategy for coping with the urbanization process (Sol et al. 2014).

With urban development, many original environmental components are cleared and replaced by human-made structures. This, in turn, leads to a prompt removal of previous habitat conditions and food items, and thus the loss of many of the organisms that depend on them (Marzluff 2001). This could be one of the reasons why most of the urban ecology research has focused on how habitat alterations and urban complexity mold urban bird diversity (Shochat et al. 2006, 2010; Lowry et al. 2013; but see Marzluff 2016 for updated topic trends). As novel types of resources become available with urban development, studies going beyond bird community emergent properties are needed to fully understand the complexity of the phenomenon. For instance, among the set of novel feeding resources, direct food provisioning by people (Galbraith et al. 2015), as well as indirect human sources (e.g., litter), which are virtually ubiquitous across urban systems, head the list (Auman et al. 2008; Crates et al. 2016). Plant material and fruits from exotic and native, often planted, vegetation components are also common in urban greenspaces (Reichard et al. 2001; Lafleur et al. 2007; Quesada and MacGregor-Fors 2010). In addition, birds can use buildings for nesting (Møller 2010a) and even as feeding sites by exploiting insects trapped by polarized light pollution (Robertson et al. 2010). The availability of such resources may facilitate the persistence and the recruitment of individuals to the new anthropogenic system, as long as organisms are able to take advantage of these opportunities and can tolerate the disturbances that characterize cities (Emlen 1974; see Chaps. 3 and 4). Thus, urban areas impose new ecological conditions that not all organisms are able to deal with (Møller 2009; Sol et al. 2014), but in which others can persist and even thrive (Sih et al. 2011). These differential

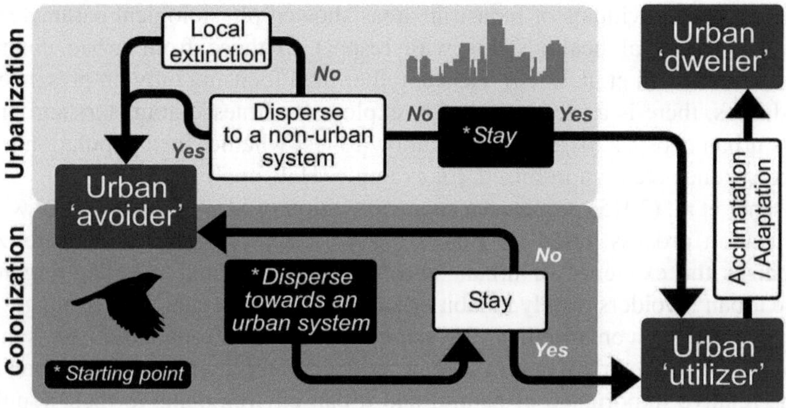

Fig. 6.1 Decision-making model depicting the chances of living in urban areas and the behavioral challenges during each state. See Sect. 6.4 (this chapter) for details on the behavioral challenges related to the response of birds to urbanization

responses can be understood by exploring the ability of species, populations, and individuals to use anthropogenic conditions in urban systems (Hulme-Beaman et al. 2016; see Sect. 6.3 in this chapter). The initial response of organisms to altered environmental conditions is probably mediated by behavior and their decisions to leave an urbanized area or stay and colonize, and in doing so, gain the opportunity to adjust to the new environmental scenario (Lowry et al. 2013; Sol et al. 2013a; Wong and Candolin 2015; Fig. 6.1).

6.2 Avian Responses to Urbanization

Back in 1996, Robert B. Blair proposed a terminology to characterize the way in which birds respond to urbanization, which has facilitated important insights into the underlying mechanisms of responses to this global environmental change (e.g., Bonier et al. 2007; Kark et al. 2007). This terminology categorized species as 'avoiders', 'adapters', or 'exploiters' based on their presence and abundance in natural, moderately, or highly urbanized areas, respectively. Despite its usefulness, Blair's terminology rules out, among other things, the fact that the density of a native species in urban environments is, to some extent, related to their density in the surrounding areas (Sol et al. 2013b, 2014). This shortcoming may hamper a fuller understanding of the mechanisms that constrain or promote avian distribution and abundances in urban areas, such as intraspecific responses to different urbanization scenarios (Evans et al. 2011; Sol et al. 2013b, 2014). In this sense, the possibility of birds having metapopulation dynamics in urban systems has also been vaguely considered (Padilla and Rodewald 2015). For instance, despite high abundances of the House Sparrow (*Passer domesticus*) in both urban and industrial areas of a

Mexican city, individuals of industrial areas showed physiological parameters that suggest detrimental health status with respect to those from urban conditions (Chávez-Zichinelli et al. 2010). Hence, rather than focusing only on presences and abundances, there is a pressing need to explore for fitness parameters and proxies across urban ecosystems, which may allow to test whether metapopulation growth and persistence are compromised, for example (Hale et al. 2015).

Fischer et al. (2015) proposed a characterization of how organisms deal with the urbanization process based on Blair's (1996) terminology. This characterization recognizes the existence of urban 'avoider', 'utilizer', and 'dweller' organisms. While urban 'avoiders' rarely inhabit urban areas, Fischer et al. (2015) suggest they may persist in such environments by using the resources from natural areas that are scattered across the urban matrix. Urban 'utilizers' and 'dwellers' are defined based on the relative importance of natural and urban environments to their population dynamics, whereby 'dwellers' are located at one extreme of this gradient of response and their persistence in urban areas is independent of extra-urban resources. Population responses to urbanization of particular species can fully fall within a particular category or span multiple categories, which recognize the dynamic nature of the adjustment of organisms to this environmental change (Sol et al. 2014; Fischer et al. 2015).

Based on the above, we believe that the study of interactions between organisms and their environments under a behavioral framework could strengthen the understanding of avian responses to urbanization. Taking into account that behavior affects key demographics parameters and population dynamics, an ethological scope could shed light on whether organisms are potential 'avoiders', 'utilizers', or 'dwellers'. In this chapter, we pinpoint the importance of avian behavior in urban areas as a key element in the study of the mechanisms that animals use to confront not only urbanization, but also other rapid human-induced environmental changes such as biological invasions, overharvesting, pollution, and climate change (Sih et al. 2011; Maxwell et al. 2016).

For Latin American cities in particular, the study of avian behavior offers unprecedented opportunities for research into the behavioral repertories of its generally understudied avifauna. This knowledge may not only help our understanding of the differential fate of organisms based on their behavioral profiles and performances, but may also encourage more predictive knowledge that, for instance, could promote better urban planning for sustainable urban development, as well as conservation strategies that are so urgently required in the region (United Nations 2015; see Chaps. 1, 2, and 8). Here, we provide an overview of avian behavioral research in urban areas, drawing from studies worldwide. We then review avian behavior studies in urban Latin America and contrast them with the findings of the behavioral literature from the rest of the world, highlighting insights, identifying potential knowledge gaps, and suggesting future directions on avian behavior research in an increasingly urbanized Latin America.

6.2.1 Ethological Adjustments to Novel Environments

The current wave of human-induced rapid environmental changes, including urbanization (Sih et al. 2011), requires appropriate adaptive responses from organisms to persist despite the new challenges. As noted above, behavior is the first tool that birds, as well as other animals, have when they confront new scenarios (Tuomainen and Candolin 2011), and it can be described and studied using a variety of approaches (e.g., descriptive, experimental, correlational). Thus, the literature vastly illustrates behavioral comparisons between urban and nonurban populations of focal species, finding striking differences among them (Candolin and Wong 2012; Gil and Brumm 2014; Wong and Candolin 2015; see Sect. 6.3 in this chapter). These changes may be the product of individual selection or a plastic response to new scenarios (Sol et al. 2013a). Yet, disentangling these two effects is fraught with difficulties as the two processes may be occurring simultaneously. On the one hand, to survive to increased urbanization intensity or to successfully establish in urbanized areas may require the selection of particular phenotypes from the original population that can 'perform well' in these environments (Carrete and Tella 2011; Miranda et al. 2013; Partecke 2014). For instance, the frequency of certain genes within a population may favor the ability to withstand urban life (Müller et al. 2013) or to establish in new areas (Edelaar et al. 2015). However, genetic selection, even though it is always active to some extent within a population (Carrete et al. 2016), may take longer and requires stable selection pressures, and not necessarily be the most parsimonious way for responding to quick changes occurring during urbanization processes (van Buskirk 2012). On the other hand, many different behavioral responses may emerge as a consequence of a particular plastic response. If we consider the distinct components of plasticity, it is important to recognize the cases in which the same genotype provokes a variety of responses to different environments. In the context of behavioral plasticity, this has been termed developmental plasticity (sensu Snell-Rood 2013) or reaction norm (Levis and Pfennig 2016).

Developmental plasticity is the range of ways that a particular behavior may express itself in a particular situation (Snell-Rood 2013); thus, some behavioral syndromes may be selected as they perform well in cities without any direct influence from genetic selection (Dingemanse and Wolf 2013). Indeed, although the House Sparrow is a species whose urban and nonurban populations differ behaviorally (Seress et al. 2011; Bókony et al. 2012), studies have failed to find any significant genetic differences between individuals from both environments (Vangestel et al. 2011). Another type of behavioral plasticity that may be observed in birds is activational plasticity (sensu Snell-Rood 2013), which refers to the variable response that an animal exhibits to the same stimulus through its life. It is clear that, after an initial experience with a certain stimulus, an organism may adjust its responses and improve its efficiency in successive encounters until it reaches an optimal response. One of the main mechanisms responsible for this intraindividual change in behavior is learning (Sol et al. 2013a), understood as the alteration of behavior as a result of individual experience (Shettleworth 2010). Birds can acquire experience through two main nonexclusive mechanisms, individual and social learning. In individual learning, acquiring new information occurs through experiences based on trial and error and successful rewards. Social learning, on the other hand, is related to information acquisition through the observation of other individuals and the consequences of their behavior (Brown and Laland 2003), which has been suggested to be a useful spreading mechanism for behavioral innovation and thus a useful skill for thriving in cities (Brown 2012). It is relevant to notice that innovative problem-solving may also be achieved by simple behavioral persistence without the need to invoke demanding cognitive processes, such as learning (Guez and Griffin 2016). Regardless of the origin of behavioral changes, the effectiveness of ethological adjustments is thought to have several properties that make them ideal for explaining many ecological and adaptive changes for the urban living. For example, behavioral adjustments can occur more quickly than any physiological or morphological changes that are mediated by genetic mechanisms; so, the time lag in the expression of the phenotype that fits the new situation is shorter (see Hendry et al. 2008 and references therein). Nonetheless, there is an additional advantage: some behavioral changes can be reversed, which is particularly useful in changing environments such as built-up systems (West-Eberhard 1989).

Despite all of the highlighted advantages, it is worth underlining that not all behavioral responses are favorable in terms of fitness and, for example, maladaptive behavioral responses may also occur when urbanization erodes the reliability of the cues that animals use to assess the quality of resources or mates (Wong and Candolin 2015, see Sect. 6.3 in this chapter). Finally, behavioral changes in urban individuals and populations must translate into an improvement or upholding of fitness so that a population can persist in the new environment. Establishing direct links between the development of a particular behavioral response and its impact on population dynamics is a difficult task (Pelletier and Garant 2012), as is whether or not behavioral adjustments are sufficiently vigorous to be able to deal human-induced environmental changes (Wong and Candolin 2015).

6.3 Behavioral Challenges in Urban Areas

Urbanization creates new environments that force birds to tackle important challenges (Shanahan et al. 2014). Among the drastic changes given by the urbanization process, situations that are stressful in nonurban areas (e.g., predators eliciting acute responses) may become commonplace and persistent in cities. These situations can turn into chronic stress, affecting the allostatic condition in birds (e.g., Canyon Towhee—*Melozone fusca*; Chávez-Zichinelli et al. 2013), thereby hampering their ability to benefit from anthropogenic resources. For instance, corticosterone plays a key role in behavioral and physiological modulation during stressful situations that could influence on whether a bird adjusts or not to the urban life (Bonier 2012). Nevertheless, no consistent patterns regarding the relationship between urban life and this hormone have yet been established (Bonier 2012; Buchanan and Partecke 2012).

In addition to the replacement of preexistent systems with artificial structures and the intrinsic urban disturbances, cities are often characterized by high levels of noise, chemical, and light pollution (Forman 2014). These urban features have been shown to influence avian physiology including the endocrine, nervous, and immune systems, which can lead to important behavioral changes (Gil and Brumm 2014). For instance, chemical pollution is an important disruptor of the endocrine and nervous systems that control many aspects of sexual and social behavior (e.g., aggressiveness, courtship). Indeed, birds subject to high concentrations of heavy metals are more aggressive (Janssens et al. 2003) and more dominant (Hogstad and Pedersen 2007), states that may affect their breeding performance (Eeva and Lehikoinen 2000; Dods et al. 2005). Chemical pollution may also affect signaling indirectly. For example, birds exposed to DDT significantly change brain structures that are related to song (Iwaniuk et al. 2006), and animals with higher mercury concentrations in their blood are known to modify their song repertoires (Mckay and Maher 2012). Artificial lighting is another of the common extended forms of pollution in urban areas that directly affect the endocrine system (Dominoni et al. 2013; Da Silva et al. 2015) and is also a good example of how cities induce changes in wildlife use of space and time (Dominoni 2015).

Probably one of the most studied sources of behavioral change in urban areas is the use of novel resources resulting from human activities (see 6.1 in this chapter). A classic example is the Great Tit (*Parus major*), which starts to open milk bottles to feed on the cream. The first observations of this technical innovation were recorded near Southampton (England). Afterward, the innovative behavior became widespread across many regions of the United Kingdom (Fisher and Hinde 1949). The mechanisms underlying the transmission of these innovations are frequently based on social learning and cultural transmission, in which social interaction plays a crucial role (Bókony et al. 2009). Since sociality is mediated by external factors such as predation risk and resource abundance, social dynamics are expected to change in cities because these external factors differ between urban and nonurban areas (Blumstein 2014). Density-dependent processes are often mediated by cost/benefit relationships and, thus, social interaction is expected to have an influence on fitness and, ultimately, on population dynamics (Blumstein 2014). Both sexual and social interactions are frequently performed via signaling, where song, integuments, and plumage usually work as cues of quality (e.g., dominance status, foraging ability, body condition). However, cues that work in original habitats may not be informative in urban systems and could even result in maladaptive behavioral responses. For example, certain urban features may compromise not only signal production but also their transmission. One of the most studied such cases is that of song transmission in noisy urban environments (Slabbekoorn and Peet 2003; Laiolo 2010; Gil and Brumm 2014). Urban areas are characterized by noise that overlaps with the song frequencies of many birds, which has driven individuals to adjust their behavior to overcome sound interference. Adjustments that may be beneficial for communicating in a noisy environment include changes in spatiotemporal activity, song frequencies, and song structures (Rosenthal and Stuart-Fox 2012); yet, these behavioral adjustments do not necessarily guarantee the ability to persist and thrive in urban areas (Moirón et al. 2015). Visual signals can also be masked in a city, and signaling traits may thus mislead messages with serious consequences for the receptor's fitness. For instance, Senar et al. (2014) showed that the black tie in Great Tits, a sexually selected trait that signals parental ability (Norris 1990), is positively related to survival in nonurban habitats; however, the relationship between survival and the size of the black tie is negative in cities. Thus, if urban females use big ties as a quality signal during mate choice (see Norris 1993, Quesada and Senar 2009), they are, in fact, eliciting a maladaptive response and selecting the least viable males, and consequently may succumb to an evolutionary trap (Senar et al. 2014).

Despite the negative impacts of urbanization, some bird species have shown to benefit from urbanized systems. For instance, tolerance to human disturbance may improve productivity as living in close proximity to humans may discourage potential predators (Møller 2010a). This may be determinant for commensal species that increase their breeding success dramatically in cities (Hulme-Beaman et al. 2016). Comfortable ecological conditions in cities may also affect bird movements like migratory behavior, such as urban populations that become more sedentary than nonurban ones (Evans et al. 2012), presumably due to the synergistic effect of the greater availability of feeding resources coupled with climatic amelioration (Plummer et al. 2015). Recent evidence has also shown that cities can influence migrant phenology, and migrants from urban areas arrive relatively earlier than their nonurban counterparts (Tryjanowski et al. 2013).

Finally, many aspects of reproduction and breeding performance may also vary between urban and nonurban environments (Chamberlain et al. 2009; Shanahan et al. 2014). For instance, highly abundant food sources in cities may potentially make urban birds more productive; yet, evidence shows that the breeding output of urban birds is lower when compared with their performance in original habitats (Chamberlain et al. 2009). Nevertheless, in San Diego (CA, USA), milder climatic conditions enabled Dark-eyed Juncos (*Junco hyemalis*) to advance and even extend their breeding phenology (a plastic adaptive response), increasing fitness and productivity in this population by changes in their reproductive behavior (Yeh and Price 2004).

6.4 Behavioral Background of 'Avoiders', 'Utilizers', and 'Dwellers'

A fundamental question that must be raised when talking about urbanization is: what leads organisms to avoid, utilize, or dwell within an urban area? Some authors have approached this problem by suggesting that the processes involved in the arrival of a species to a city resemble those than occur when exotic species invade new environments (Blackburn et al. 2009; Evans et al. 2010; Sol et al. 2013a; Møller

2014). From a behavioral point of view, we can examine this question from many different perspectives, all of which are related to the final decision taken by the animal: 'stay' or 'leave'. The choice must probably be based on a fitness cost–benefit analysis (Chaine and Clobert 2012). If 'stay' is the response, organisms must deal with the new environmental challenges and maximize their fitness (e.g., survival, reproduction), considering factors that differ from their natural environments such as food, mates, nesting sites, predators, humans, parasites, and the disruption of communication, among others (Candolin and Wong 2012; Gil and Brumm 2014). However, there are certain scenarios where, despite that birds decide to stay or colonize urban areas, they may show negative population growth rates, as happens in source–sink dynamics (Padilla and Rodewald 2015; see Sect. 6.2 in this chapter).

Adjustments via behavioral responses and adaptation through genetic changes are two possible strategies for accommodating, stabilizing, or improving population status (i.e., growth in numbers; Evans et al. 2010; Chaine and Clobert 2012; Sol et al. 2013a). Consequently, the first step for assessing avian responses to urban life is to understand if birds leave their original natural habitats and disperse into built-up areas (Evans et al. 2010; Sol et al. 2013a; Møller 2014), or whether resident species avoid cities as a way to compensate for the negative impact of built-up environments where some important ecological requirements may no longer be fully satisfied (Biamonte et al. 2011; Chaine and Clobert 2012). Once the urban environment has been reached, or once the system has been urbanized, individuals must evaluate the feasibility of staying (Fig. 6.1). As has been noted by Fischer et al. (2015), there is a possibility that some resident species will decide to stay in patches with original habitats or novel systems that resemble them across cities, which may still satisfy their ecological requirements. However, individuals that stay ought be able to withstand urban-related perturbations through adjustments, such as their flight initiation distances (Møller et al. 2013) and home range sizes (Chiang et al. 2012). Hence, we hypothesize that for such individuals, behavioral responses may be much more important than genetic changes in the short and medium terms.

In scenarios under which individuals are urban 'utilizers' or 'dwellers', their ability to 'choose' among conditions will be pivotal because it cannot depend anymore, or at least not entirely, on nonurban resources. In fact, behavioral plasticity and genetic changes are probably more important in these cases, with a crucial role on bird ability to use anthropogenic niches (Hulme-Beaman et al. 2016). Thus, birds unable to choose among the array of available conditions will probably become urban 'avoiders' from an ethological perspective (Hulme-Beaman et al. 2016; Sih et al. 2016; Fig. 6.1). Recognizing the existence of this intraspecific variation between 'avoiders' and 'dwellers' may shed important light on identifying and understanding the influence of demographic parameters driving urban bird ecology.

Despite the existence of variations between populations in response to urbanization, these are lower than those observed between species, which supports the idea that the ability to thrive in cities is a species-level trait (Sol et al. 2014). Indeed, some traits have been identified as being more closely associated with avian species inhabiting cities, including: high dispersal ability (Møller 2009), low propensity to migrate (Croci et al. 2008), and high fecundity (Møller 2009). However, fitting into one or more of the aforementioned traits does not automatically turn a bird species able to become an urban 'utilizer' or 'dweller'. Adaptive traits may only be pivotal if they generate a positive impact on population dynamics (Pelletier and Garant 2012; Wong and Candolin 2015). The population size of 'incomers' facing urban challenges is usually small and determined by environmental stochasticity, which may easily lead to local extinction. Indeed, population abundance in surrounding nonurban environments can directly be related to species abundance in urban areas, suggesting that some species that are present in urban areas are there due to random dispersal (Sol et al. 2014), and would also be classified as urban 'utilizers'.

After their initial arrival to the city, some species increase their numbers dramatically in comparison with the nonurban system in which they inhabit, a process whose underlying mechanisms are still subject to debate (Møller 2009; Evans et al. 2011; Jokimaki et al. 2014). Studies that take into account the influence of the population size in surrounding nonurban environments have shown that the association of some traits previously related to the presence of avian species in cities provide little support as drivers of differential avian success in urban areas (Evans et al. 2011; Sol et al. 2014). Likewise, larger brains and associated cognitive abilities turn out to be less important (Evans et al. 2011; Møller and Erritzøe 2015). These findings do not mean that behavioral flexibility is not relevant for a life in the city; rather, a number of different strategies including behavioral flexibility may assist when it comes to thriving in an urban environment (Sol et al. 2014).

6.5 Avian Behavior in Urban Latin America

To summarize the literature that covers several dimensions of avian ecology in urban Latin America and implicitly or explicitly discuss bird behavior, we conducted a bibliographic search in the Web of Science platform (www.webofknowledge.com) and Google Scholar (http://scholar.google.com) seeking for peer-reviewed publications, theses, conferences, and congress proceedings, as well as magazines published by local universities and ONGs. We used a set of keyword combinations, both in Spanish and English, including the terms 'urban', 'urbanization', 'city', 'bird', and 'avian', with refining keywords, such as 'behavior', 'behavioral syndrome', 'personality', 'risk-taking', 'temperament', 'copying styles', 'behavioral flexibility', and 'behavioral plasticity'. From the list of articles obtained with the combinations using the aforementioned keywords, we selected those that were performed in urbanized areas from Latin America. We retreived a total of 86 publications from 11 countries that met our search criteria including additional publications from the references cited in some of the gathered documents (Table 6.1). To analyze the gathered literature, we explored the role of avian behavior when it comes to confront some of the main challenges to persist or colonize urban areas in Latin America (see Shanahan et al. 2014). Specifically, we focus on five types of urban

Table 6.1 Behavioral studies from Latin America addressing the different challenges for an urban life

Topic[a]	Countries	Experimental (%)	Observational (%)
Habitat loss and fragmentation (19)	Argentina, Brazil, Chile, Colombia, Costa Rica, Ecuador, Mexico	16	84
Resources availability (24)	Argentina, Barbados, Brazil, Chile, Colombia Mexico, Peru, Venezuela	33	67
Pollution (14)	Argentina, Brazil, Chile, Colombia, Costa Rica, Mexico	71	29
Species interaction (9)	Argentina, Bolivia, Mexico, Peru, Venezuela	44	66
Human activities (20)	Argentina, Barbados, Brazil, Colombia, Mexico	40	60

[a]Number of studies of each topic in parenthesis

challenges: (i) habitat loss and fragmentation; (ii) resource availability; (iii) pollution; (iv) species interactions; and (v) human activities.

6.5.1 Habitat Loss and Fragmentation

Under the context of the urbanization process, examples focused on the behavioral adjustments of vertebrates, including birds, to habitat loss and fragmentation are scarce (Lowry et al. 2013; Sol et al. 2013a). Although behavior adjustment has been recorded in relation to human disturbances (Salinas-Melgoza et al. 2013), we did not find any study directly assessing avian behavioral adjustments to habitat loss and fragmentation in urban Latin America. Yet, we found studies that suggest that some native species take advantage of these urban conditions, such as the Rufous-tailed Hummingbird (*Amazila tzacatl*) in Costa Rica (Biamonte et al. 2011), Rufous-collared Sparrow (*Zonotrichia capensis*) and Monk Parakeet (*Myiopsitta monachus*) in Argentina (Bellocq et al. 2011; Bucher and Aramburú 2014; see Chaps. 2 and 3). Despite the high abundances of these species, whether they persist, thrive, and finally become urban 'dwellers' in the long term remains unknown (Sol et al. 2013a; Wong and Candolin 2015). Indeed, individuals from urban populations of the Rufous-collared Sparrow—which are highly abundant in some Chilean (Ruiz et al. 2002) and Argentinean urban centers (Bellocq et al. 2011)—have lower body weights, and several of their physiological blood parameters suggest they live in more stressful environments than their nonurban counterparts, as has been recorded in central Chile (e.g., Ruiz et al. 2002). Accordingly, the Common Ground-Dove (*Columbina passerina*) was highly abundant in gardens and open areas from San Jose in Costa Rica up to the late 1990s, but nowadays has almost disappeared from urban areas from this region (Biamonte et al. 2011). These findings reinforce the need to assess not only species presence and abundance in urban areas but also their behavioral responses and whether or not they are sufficient to confront the habitat change produced by urbanization.

6.5.2 Resource Availability

Based on the retrieved studies, we identified at least two ways of how birds use resources available in cities. First, in most of the reviewed studies, species use food items that are not directly generated by anthropogenic activities (referred to as natural food hereafter) that are mostly present in urban greenspaces. For instance, the presence of patches of native vegetation scattered throughout Brazilian urban areas allow the presence of plant–frugivorous mutualistic networks with birds remaining as important seed dispersers (Andrade et al. 2011). A study performed in Mexico indicates that Nearctic–Neotropical migrant birds may also make use of natural food resources and shelter in urban greenspaces, emphasizing the importance of greenspaces in cities (Charre 2013). Other examples of species using natural food items in urban areas include the Yellow Warbler (*Setophaga petechia*) and Summer Tanager (*Piranga rubra*) in Colombia (Caicedo-Argüelles and Cruz-Bernate 2014), seven species of hummingbirds and two passerines in a university campus in Brazil (Mendoça and dos Anjos 2006), and a terrestrial bird assemblage from an urban park in Venezuela (Sainz-Borgo 2016), among others. Not just food, but nesting sites on natural components scattered across the urban area are also chosen by birds. For example, Zuria and Rendón-Hernández (2010) followed nests of six bird species in a greenspace of the city of Pachuca (Mexico), recording differential substrate preferences (i.e., native, exotic vegetation).

Other examples regarding birds nesting in urban areas include those of the Rufous-tailed Hummingbird (*Amazila tzacatl*), Bananaquit (*Coereba flaveola*), and Yellow-headed Caracara (*Milvago chimachima*), among many other Colombian species (Borrero 1965; De La Ossa and De La Ossa-Lacayo 2011). The descriptive nature of the above studies precludes us to categorize these urban birds as 'avoiders' or 'dwellers', as there is no available information regarding their behavioral decisions. This is not surprising as most of the reviewed studies were observational (see Table 6.1). Additionally, some species have been identified to have the ability to change their usual foraging strategies and hence able to incorporate vegetable material or fruit from exotic trees to their diets, or take profit from artificial food items such as food waste or direct feeding by people. These behavioral adjustments have been more commonly recorded in harsh environmental conditions. For instance, during winter, several fruit-eating bird species from Brazil, such as the Rufous-bellied Thrush (*Turdus rufiventris*), Pale-breasted Thrush (*Turdus leucomelas*), Purple-throated Euphonia (*Euphonia chlorotica*), among others, change their movement patterns and visit urban areas for feeding on fleshy fruits of nonindigenous species from urban greenspaces, but then return to their natural areas to breed (Guix

2007). A similar situation has been described for the Black-chinned Siskin (*Carduelis barbata*) in Chile, where individuals of this species have been recorded feeding on seeds from exotic trees and using them as refuges in built-up areas in winter (San Martín 2009). Other examples of the ability to choose nonhabitual resources in an urban context include the nest built on an open-sky telephone by Azure-crowned Hummingbirds (*Amazilia cyanocephala*) in Mexico (Escobar-Ibáñez and MacGregor-Fors 2015) and the nectarivorous behavior of House Sparrows as function of temporal variability in human disturbance in Argentina (Leveau 2008). Both behavioral plasticity and personality traits may help to explain their ability to use novel urban resources in Latin America. For instance, a field study using experimental feeders in an extra-urban area in Argentina showed that granivorous birds not related to urbanization did not use the feeders, regardless of their feeding preference, underlining their aversion to novel situations (e.g., neophobia; Echeverría et al. 2006). Likewise, another experimental study performed in Argentina showed that the Chimango Caracara (*Milvago chimango*) is characterized by its tendency to explore, showing lack of neophobia and great ability to innovate (Biondi et al. 2010). These personality traits and learning skills give this species the impressive capacity to obtain food in novel situations by opening garbage bags and raiding trash cans (Biondi et al. 2008; CG-L pers. obs.; Fig. 6.2). Even within species, different levels of neophobia have been found in relation to a bird's original habitat. Related to the latter, Ducatez et al. (2016) conducted field experiments on nine avian species in Barbados and found that urban birds are bolder, less neophobic, and have shorter flight distances than their less urbanized conspecifics. Even so, less neophobia in urban birds is not a consistent pattern and, unexpectedly, clear generalist and urban species such as the House Sparrow, Eared Dove (*Zenaida auriculata*), and Shiny Cowbird (*Molothrus bonariensis*) are usually neophobic (Echeverría et al. 2006; Echeverría and Vassallo 2008). These differences have also been recorded within species, such as the study of Audet et al. (2016), who found that urban Barbados Bullfinches (*Loxigilla barbadensis*) are highly neophobic, but bolder and better at problem-solving than nonurban individuals. Discrepancies in the degree of neophobia in birds that usually inhabit urban areas may be due to the modification of fear, which is directed to regularly encounter objects rather than a general phenotypic characteristic to deal with the urban habitat (Greggor et al. 2016).

6.5.3 Pollution

With the exception of two studies focused on artificial night lighting in urban Latin America (i.e., MacGregor-Fors et al. 2011; Dorado-Correa et al. 2016), noise has been the most studied type of pollution affecting birds (Bermúdez-Cuamatzin et al. 2011). The reviewed studies focused on noise have revealed several ways in which birds adjust their songs to urban life. For example, some species modify the frequency of a specific syllable (Bermúdez-Cuamatzin et al. 2011) or combine changes

Fig. 6.2 Chimango Caracara (*Milvago chimango*) exploring on garbage bags (*above*) and group of individuals of the same species feeding on leftovers in an asphalt street (*below*) (Photo: CG-L)

in the length of songs with changes in frequencies (León et al. 2014; Pacheco-Vargas and Losada-Prado 2015; Soto-Gamboa et al. 2015). Some studies also provide evidence of individual song plasticity (Bermúdez-Cuamatzin et al. 2009; Ríos-Chelén et al. 2013). Of all the types of song modification, changes in frequencies are the most studied behavioral responses to noise. For instance, by considering the variation present in five frequency parameters, Laiolo (2011a) showed that Rufous-collared Sparrows sang significantly higher pitched songs (with higher minimum frequencies) in noisy than in quiet areas in Bahía Blanca, Argentina. A similar pattern was recorded for this species in Chile (Soto-Gamboa et al. 2015), as well as for other species, such as the House Finch (*Haemorhous mexicanus*) in Mexico (Bermúdez-Cuamatzin et al. 2009; Bermúdez-Cuamatzin et al. 2011), House Wren (*Troglodytes aedon*) in Costa Rica (Redondo et al. 2013), Saffron Finch (*Sicalis flaveola*) in Argentina (León et al. 2014), and Rufous-browed Peppershrike (*Cyclarhis gujanensis*) and Scrub Greenlet (*Hylophilus flavipes*) in Colombia (Pacheco-Vargas and Losada-Prado 2015).

The relationship between noise and shifts in song frequency showed to be stronger in oscines than in suboscines from Mexico and Brazil, showing that suboscines may be more vulnerable to acoustic pollution than oscines and thus less capable of entering cities, as well as to acoustically novel habitats (Bermúdez-Cuamatzin et al. 2009). In terms of variability, a study from Argentina revealed that neighboring Rufous-collared Sparrows in gardens sang similarly when compared with those from nonurban environments, thereby suggesting that the filtering and homogenization of singing behavior can occur in urban areas (Laiolo 2011b).

However, another study of the same species in Chile found great individual variability in repertoire size and song cues in urban environments (Soto-Gamboa et al. 2015). It is possible that these differences are given by different colonization stages, with the reduction in repertoire size at initial stages and subsequent increases in variability (see Møller 2010b for a similar example but for avian fear responses); however, methodologies could also lead to such differences. Finally, a recent study on the same species explored why urban individuals start singing earlier than individuals from nonurban areas. The authors found that noise—but not lighting levels—determined variations in this behavior in Rufous-collared Sparrows in Bogota, Colombia (Dorado-Correa et al. 2016).

6.5.4 *Species Interactions*

Studies focused on avian behavioral changes related to competition for resources, predation, and parasitism were scattered, but present, in urban Latin America. For example, the increase in abundance of some potential preys in cities has been recorded to be beneficial for avian predators such as the Harris's Hawk (*Parabuteo unicinctus*) that feed on bats in Mexico City (Ortega-Álvarez and Calderón-Parra 2014), as well as Peregrine Falcons (*Falco peregrinus*) recorded preying on the highly abundant exotic and invasive Monk Parakeets in Santiago de Chile (Fig. 6.3). Supplementary food in urban centers may also determine hierarchical relationship among birds, as reported in Venezuela, where a nonrandom order of access to an artificial food patch was recorded for several avian species (Levin et al. 2000). This type of subsidized urban food may also alter bird–plant interactions. For instance, two studies in Mexico show that the presence of nectar feeders decreases plant visitation rates of all studied hummingbirds (Arizmendi et al. 2007, 2008). Behavioral shifts in the urban context may also affect the relationship between parasites and their hosts, as has been shown in Mexico where two widely distributed urban birds started to use cigarette butts to build their nests, a behavior that was found to be associated with a decrease in the number of nest-dwelling parasites (Suárez-Rodríguez et al. 2012, see Chap. 5). A recent study has reported that despite this behavior decreases parasite load on nests, it might increase breeding costs for urban birds as a consequence of the genotoxic damage produced by cigarette butts (Suárez-Rodríguez et al. 2017). Finally, as recorded in other parts of the world (Shanahan et al. 2014), results of nest predation along urbanization gradients are contradictory. For example, a study from Mexico reveals that nest survivorship was higher in nonurban than in urbanized environments; nevertheless, the speed at which nests were depredated did not change along the gradient of urbanization intensity (López-Flores et al. 2009). By contrast, a study conducted in Bolivia showed that the predation and destruction of artificial nests did not differ between urban and nonurban sites (Salazar and Garitano-Zavala 2015). These contrasting observations maybe related to the findings of Rivera-López and MacGregor-Fors (2016), who showed that artificial nest predation was driven by the interaction between spatial location

Fig. 6.3 Peregrine Falcon (*Falco peregrinus*) preying on an invasive Monk Parakeet (*Myiopsitta monachus*) in Santiago de Chile (Photo: Romina Benzi)

and degree of urban development, where visual predators were most frequent. This may help to explain contrasting results between studies in terms of nest predation, the dynamics of which can be diverse within the same urban area (both predation 'relaxation' and intensification within the same area; Rivera-López and MacGregor-Fors 2016).

6.5.5 Human Activities

The increase of human population density and high environmental disturbance regimes in urban areas are other challenges shown to drive bird behavior in urban Latin America. In fact, compared with their nonurban conspecifics, urban birds of several species from southeast Brazil have shown higher levels of tolerance to human disturbance and shorter flight initiation distances (Piratelli et al. 2015). The same pattern has been recorded for Burrowing Owls (*Athene cunicularia*; Rebolo-Ifrán et al. 2015; Cavalli et al. 2016). Variability in flight initiation distances has also been associated with the number of predators that approach the implied birds (Piratelli et al. 2015; Cavalli et al. 2016), as well as their body mass (Piratelli et al. 2015), but has not been related with predator type (Fabiano et al. 2015). Interindividual variability in flight initiation distances and the relative brain size of a given species could be good predictors of the ability to colonize a city, as indicated for 42 avian species from Argentina (Carrete and Tella 2011). Despite that habituation has been claimed as the main mechanism behind the above-described patterns (Møller 2008, 2010b), studies focused on the Burrowing Owl in Argentina found high consistency in the fear of humans throughout the adult life span of both urban and nonurban birds (Carrete and Tella 2016). This supports the existence of heritability or resemblances between relatives and that selection pressures on flight initiation distances may be an important evolutionary force in this species when confronting urbanization (Carrete et al. 2016). Moreover, a study focused on the potential physiological mechanisms behind flight initiation distances found no support that corticosterone, a glucocorticoid related to stress responses in birds (Wingfield et al. 1998), drives such behavioral response by the Burrowing Owl in Argentina (Rebolo-Ifrán et al. 2015).

6.6 Conclusions, Research Gaps, and Future Directions

Given that behavior is one of the most immediate skills available to confront environmental challenges and it influences several aspects of animal ecology, avian urban ecology will undoubtedly be strengthened by the consolidation of a behavioral approach. However, most of the reviewed studies from urban Latin America are observational, with the behavioral component of avian responses nonexplicitly evaluated. For instance, many of the reviewed studies report an increase in the abundance of some species following the loss of natural habitat by urbanization, a common pattern worldwide (see Aronson et al. 2014; Sol et al. 2014). Due to the lack of direct assessments of behavioral responses under a more mechanistic framework, we can only speculate either behavioral adjustments or the filtering of specific phenotypes underly the apparent success of some urban birds. Additionally, a common and unsurprising tendency was that several avian species across the region are using resources from remnant patches of natural habitat within the urban matrix. On the one hand, this suggests that the presence of birds in urban areas is supported by this type of resources. Nevertheless, whether or not birds are dependent on these resources, and thus fall on the 'avoider' or 'utilizer' categories, remain as open questions. The study of bird behavior in urban greenspaces also represents a bias in Latin America (Chap. 9). Regarding studies focused on behavioral changes between urban and nonurban populations, we found experimental studies showing the importance of specific personality traits in foraging adjustments that agree with findings from research performed across the globe. Experimental studies were also targeted to evaluate how birds confront the challenge of urban pollution, recording a wide array of singing adjustments, also in agreement with the global literature (reviewed by Sol et al. 2013a). Among the identified research gaps, the impact of behavioral adjustments on population dynamics has not been evaluated in Latin America (nor has been assessed in other regions; Marzluff 2016). Moreover, whether metapopulation dynamics are involved in the persistence of avian populations in urban areas is unknown, despite representing critical information to generate conservation strategies (Padilla and Rodewald 2015).

To advance our knowledge regarding the urban drivers and consequences of urbanization on birds, we must identify the role of behavioral responses to determine the varying fate of avian species, populations, and individuals confronting the urbanization process. To do so, studies evaluating the response in areas with different urbanization intensities, ideally throughout urban systems (see Chap. 9 for a detailed analysis on citywide surveys), are crucial. Specifically, we recognize the importance of understanding the interactions of birds with anthropogenic resources with the main goal of disentangling their relationships with the urban scenarios and their related components. After identifying the behavioral responses of birds to urbanization, we ought to assess their impacts on population dynamics, unraveling whether such adjustments are advantageous or maladaptive in terms of fitness (Sol et al. 2013a; Fischer et al. 2015; Wong and Candolin 2015).

Among the approaches to study avian responses to urbanization, behavioral and evolutionary frameworks allow to assess how birds thrive or fail in these novel environments. In particular, the behavioral framework is undoubtedly more plastic than morphological or physiological ones, leading to a wide spectrum of evolutionary inquiries for which birds are excellent models. Behavioral modifications in urban areas are thought to occur under selective regimes, and so may promote adaptive changes involving genetic, physiological, and morphological modifications in future generations (Bateson and Laland 2013). However, whether or not behavior promotes evolution in urban systems is still subject to debate since behavioral plasticity may also buffer evolutionary change (Duckworth 2009). Due to its flexibility and, in some cases, reversibility, behavior is not subject to the same degree of evolutionary pressure as morphological or physiological traits (van Buskirk 2012, Rosenthal and Stuart-Fox 2012). Behavioral changes without genetic transmission may be selected, so that some phenotypes are directed toward an adaptive peak (Lopez-Sepulcre and Kokko 2012). Thus, transmitted learned behaviors may eventually become genetically stabilized in subsequent generations (e.g., Baldwin effect; Badyaev 2009). It is noticeable that these topics represent important research challenges not only in Latin America, but globally (Evans et al. 2015; Marzluff 2016). Untangling the role that behavior has on birds in urban Latin America may not only contribute to the design of conservation strategies and a better urban planning, but also to reveal basic, and still understudied, questions related to animal behavior and evolution in general.

Acknowledgments The authors are grateful to Daniel Sol, Olga Barbosa, Oriol Lapiedra, and Maria Moirón for fruitful discussions over the past years. Additionally, they also thank Romina Benzi and Fernando González for Fig. 6.3, the reviewer and the editors for their valuable comments. CG-L was supported by FONDECYT (No. 11160271) from CONICYT, Chilean government. JQ was supported by the PASSERCAT project from the Natural History Museum of Barcelona and by the CGL2016-79568-C3-3-P project from the Spanish Ministry of Economy, Industry, and Competitiveness.

References

Andrade PC, Mota JVL, Freire de Carvalho AA (2011) Interações mutualísticas entre aves frugívoras e plantas em um fragmento urbano de Mata Atlântica, Salvador, BA. Rev Bras Ornitol 19:63–73

Arizmendi M, Monterrubio-Solís C, Juárez L et al (2007) Effect of nectar feeders over diversity and abundance of hummingbirds and breeding success of two plant species in a sub-urban park next to Mexico City. Biol Conserv 136:155–158

Arizmendi M, López-Saut E, Monterrubio-Solís C et al (2008) Efecto de la presencia de bebederos artificiales sobre la diversidad y abundancia de los colibríes y el éxito reproductivo de dos especies de plantas en un parque suburbano de la ciudad de México. Ornitol Neotrop 19:491–500

Aronson MF, La Sorte FA, Nilon CH et al (2014) A global analysis of the impacts of urbanization on bird and plant diversity reveals key anthropogenic drivers. P Roy Soc Lond B Bio 281:20133330

Audet J, Ducatez S, Lefevbre L (2016) The town bird and the country bird: problem solving and immunocompetence vary with urbanization. Behav Ecol 27:637–644

Auman HJ, Meathrel CE, Richardson A (2008) Supersize me: does anthropogenic food change the body condition of Silver Gulls? A comparison between urbanized and remote, non-urbanized areas. Waterbirds 31:122–126

Badyaev AV (2009) Evolutionary significance of phenotypic accommodation in novel environments: an empirical test of the Baldwin effect. Philos Trans R Soc B 364:1125–1141

Bateson P, Laland KN (2013) Tinbergen's four questions: an appreciation and an update. Trends Ecol Evol 28:712–718

Bellocq MI, Filloy J, Gustavo AZ et al (2011) Responses in the abundance of generalist birds to environmental gradients: the Rufous-collared Sparrow (*Zonotrichia capensis*) in the southern Neotropics. Ecoscience 18:354–362

Bermúdez-Cuamatzin E, Ríos-Chelén A, Gil D et al (2009) Strategies of song adaptation to urban noise in the house finch: syllable pitch plasticity or differential syllable use? Behaviour 146:1269–1286

Bermúdez-Cuamatzin E, Ríos-Chelén A, Gil D et al (2011) Experimental evidence for real-time song frequency shift in response to urban noise in a passerine bird. Biol Lett 7:36–38

Biamonte E, Sandoval L, Chacón E et al (2011) Effect of urbanization on the avifauna in a tropical metropolitan area. Landsc Ecol 26:183–194

Biondi L, Bó MS, Vassallo AI (2008) Experimental assessment of problem solving by *Milvago chimango* (Aves: Falconiformes). J Ethol 26:113–118

Biondi L, Bó MS, Vassallo AI (2010) Inter-individual and age differences in exploration, neophobia and problem-solving ability in a Neotropical raptor (*Milvago chimango*). Anim Cogn 13:701–710

Blackburn TM, Lockwood JL, Cassey P (2009) Avian invasions: the ecology and evolution of exotic birds. Oxford University Press, Oxford

Blair RB (1996) Land use and avian species diversity along an urban gradient. Ecol Appl 6:506–519

Blumstein D (2014) Social behaviour. In: Candolin U, Wong BB (eds) Behavioural responses to a changing world: mechanisms and consequences. Oxford University Press, Oxford, pp 119–128

Bókony V, Lendvai A, Liker A et al (2009) Stress response and the value of reproduction: are birds prudent parents? Am Nat 173:589–598

Bókony V, Kulcsár A, Tóth Z et al (2012) Personality traits and behavioral syndromes in differently urbanized populations of house sparrows (*Passer domesticus*). PLoS One 7:e36639

Bonier F (2012) Hormones in the city: endocrine ecology of urban birds. Horm Behav 61:763–772

Bonier F, Martin PR, Wingfield JC (2007) Urban birds have broader environmental tolerance. Biol Lett 3:670–673

Borrero H (1965) Notas sobre el comportamiento del colibrí coli-rojo (*Amazilia tzacatl*) y el mielero (*Coereba flaveola*), en Colombia. El Hornero (B. Aires) 10:247–250

Brown C (2012) Experience and learning in changing environments. In: Candolin U, Wong BB (eds) Behavioural responses to a changing world: mechanisms and consequences. Oxford University Press, Oxford, pp 46–60
Brown C, Laland KN (2003) Social learning in fishes: a review. Fish Fish 4:280–288
Buchanan KL, Partecke J (2012) The endocrine system: can homeostasis be maintained in a changing world. In: Candolin U, Wong BB (eds) Behavioural responses to a changing world: mechanisms and consequences. Oxford University Press, Oxford, pp 159–174
Bucher EH, Aramburú RM (2014) Land-use changes and Monk Parakeet expansion in the Pampas grasslands of Argentina. J Biogeogr 41:1160–1170
van Buskirk J (2012) Behavioural plasticity and environmental change. In: Candolin U, Wong BBM (eds) Behavioural responses to a changing world: mechanisms and consequences. Oxford University Press, Oxford, pp 145–158
Caicedo-Argüelles A, Cruz-Bernate L (2014) Actividades diarias y uso de hábitat de la reinita amarilla (*Setophaga petechia*) y la piranga roja (*Piranga rubra*) en un área verde urbana de Cali, Colombia. Ornitol Neotrop 25:247–260
Candolin U, Wong BBM (eds) (2012) Behavioural responses to a changing world: mechanisms and consequences. Oxford University Press, Oxford
Carrete M, Tella J (2011) Inter-individual variability in fear of humans and relative brain size of the species are related to contemporary urban invasion in birds. PLoS One 6:e18859
Carrete M, Tella J (2016) High individual consistency in fear of humans throughout the adult lifespan of rural and urban Burrowing Owls. Sci Rep 3:3524
Carrete M, Martínez-Padilla J, Rodríguez-Martínez S et al (2016) Heritability of fear of humans in urban and rural populations of a bird species. Sci Rep 6:31060
Cavalli M, Baladrón A, Isacch JP et al (2016) Differential risk perception of rural and urban Burrowing Owls exposed to humans and dogs. Behav Process 124:60–65
Chaine AS, Clobert J (2012) Dispersal. In: Candolin U, Wong BBM (eds) Behavioural responses to a changing world: mechanisms and consequences. Oxford University Press, Oxford, pp 63–79
Chamberlain DE, Cannon AR, Toms MP et al (2009) Avian productivity in urban landscapes: a review and meta-analysis. Ibis 151:1–18
Charre GM (2013) Distribución, cambios en algunas pautas conductuales y preferencias de forrajeo de aves migratorias insectívoras en parques de la ciudad de México. Universidad Autónoma Metropolitana, México, PhD Dissertation
Chávez-Zichinelli CA, MacGregor-Fors I, Rohana PT et al (2010) Stress responses of the House Sparrow (*Passer domesticus*) to different urban land uses. Landsc Urban Plan 98:183–189
Chávez-Zichinelli CA, MacGregor-Fors I, Quesada J et al (2013) How stressed are birds in an urbanizing landscape? Relationships between the physiology of birds and three levels of habitat alteration. Condor 115:84–92
Chiang SN, Bloom PH, Bartuszevige AM et al (2012) Home range and habitat use of Cooper's Hawks in urban and natural areas. In: Lepczyk CA, Warren P (eds) Urban bird ecology and conservation: studies in avian biology. University of California Press, Los Angeles, Online content, pp 1–16
Crates RA, Firth JA, Farine DR et al (2016) Individual variation in winter supplementary food consumption and its consequences for reproduction in wild birds. J Avian Biol 5:678–689
Croci S, Butet A, Clergeau P (2008) Does urbanization filter birds on the basis of their biological traits. Condor 110:223–240
Da Silva A, Valcu M, Kempenaers B (2015) Light pollution alters the phenology of dawn and dusk singing in common European songbirds. Philos Trans R Soc B 370:20140126
De La Ossa J, De La Ossa-Lacayo A (2011) Aspectos de la densidad poblacional e historia natural de *Milvago chimachima* (AVES: Falconidae) en el área urbana de Sincelejo (Sucre, Colombia). Univ Sci Bogota 16:63–69
Dingemanse NJ, Wolf M (2013) Between-individual differences in behavioural plasticity within populations: causes and consequences. Anim Behav 85:1031–1039

Dods PL, Birmingham EM, Williams TD et al (2005) Reproductive success and contaminants in Tree Swallows (*Tachycineta bicolor*) breeding at a wastewater treatment plant. Environ Toxicol Chem 24:3106–3112

Dominoni DM (2015) The effects of light pollution on biological rhythms of birds: an integrated, mechanistic perspective. J Ornithol 156:409–418

Dominoni DM, Goymann W, Helm B et al (2013) Urban-like night illumination reduces melatonin release in European Blackbirds (*Turdus merula*): implications of city life for biological time-keeping of songbirds. Front Zool 10:60

Dorado-Correa A, Rodríguez-Rocha M, Brumm H (2016) Anthropogenic noise, but not artificial light levels predicts song behaviour in an equatorial bird. R Soc Open Sci 3:160231

Ducatez S, Audet J, Ros R et al (2016) Innovativeness and the effects of urbanization on risk-taking behaviors in wild Barbados birds. Anim Cogn 20:33–42

Duckworth RA (2009) The role of behavior in evolution: a search for mechanism. Evol Ecol 23:513–531

Echeverría AI, Vassallo AI (2008) Novelty responses in a bird assemblage inhabiting an urban area. Ethology 114:616–624

Echeverría AI, Vassallo AI, Isacch JP (2006) Experimental analysis of novelty responses in a bird assemblage inhabiting a suburban marsh. Can J Zool 84:974–980

Edelaar P, Roques S, Hobson EA et al (2015) Shared genetic diversity across the global invasive range of the monk parakeet suggests a common restricted geographic origin and the possibility of convergent selection. Mol Ecol 24:2164–2176

Eeva T, Lehikoinen E (2000) Pollution: recovery of breeding success in wild birds. Nature 403:851–852

Emlen JT (1974) An urban bird community in Tucson, Arizona: derivation, structure, regulation. Condor 76:184–197

Escobar-Ibáñez JF, MacGregor-Fors I (2015) On the tightrope: use of open sky urban telephone wires by Azure-crowned Hummingbirds (*Amazilia cyanocephala*) for nesting. Wilson J Ornithol 127:297–302

Evans KL, Hatchwell BJ, Parnell M et al (2010) A conceptual framework for the colonisation of urban areas: the blackbird *Turdus merula* as a case study. Biol Rev 85:643–667

Evans KL, Chamberlain DE, Hatchwell BJ et al (2011) What makes an urban bird? Glob Chang Biol 17:32–44

Evans KL, Newton J, Gaston KJ et al (2012) Colonisation of urban environments is associated with reduced migratory behaviour, facilitating divergence from ancestral populations. Oikos 121:634–640

Evans BS, Ryder TB, Reitsma R et al (2015) Characterizing avian survival along a rural-to-urban land use gradient. Ecology 96:1631–1640

Fabiano K, Silva J, Leal-Marques R et al (2015) Comportamento de defesa de *Athene cunicularia* (Aves: Strigiformes) em resposta a aproximação de predadores a sua toca. Paper presented on the X Neotropical Ornithological Congress and XXII Congresso Brasileiro de Ornitologia, Manaus, 19–24 July 2015

Fischer J, Schneider SC, Ahlers AA et al (2015) Categorizing wildlife responses to urbanization and conservation implications of terminology. Conserv Biol 29:1246–1248

Fisher J, Hinde RA (1949) The opening of milk bottles by birds. Br Birds 11:347–357

Forman R (2014) Urban ecology: the science of cities. Cambridge University Press, Cambridge

Galbraith JA, Beggs JR, Jones DN (2015) Supplementary feeding restructures urban bird communities. Proc Natl Acad Sci U S A 120:E2648–E2657

Gil D, Brumm H (eds) (2014) Avian urban ecology: behavioural and physiological adaptations. Oxford University Press, Oxford

Greggor A, Clayton N, Fulford A et al (2016) Street smart: faster approach towards litter in urban areas by highly neophobic corvids and less fearful birds. Anim Behav 117:123–133

Guez D, Griffin AS (2016) Unraveling the key to innovative problem solving: a test of learning versus persistence. Behav Ecol 27:1449–1460

Guix JC (2007) The role of alien plants in the composition of fruit-eating bird assemblages in Brazilian urban ecosystems. Orsis 22:87–104

Hale R, Treml E, Swearer S (2015) Evaluating the metapopulation consequences of ecological traps. Proc R Soc Lond B Biol Sci 282:20142930

Hendry AP, Farrugia TJ, Kinnison MT (2008) Human influences on rates of phenotypic change in wild animal populations. Mol Ecol 17:20–29

Hogstad O, Pedersen HC (2007) Heavy metal load and dominance hierarchy in juvenile Willow Tits during winter. Environ Pollut 147:245–247

Hulme-Beaman A, Dobney K, Cucchi T et al (2016) An ecological and evolutionary framework for commensalism in anthropogenic environments. Trends Ecol Evol 31:633–645

Iwaniuk AN, Koperski DT, Cheng KM et al (2006) The effects of environmental exposure to DDT on the brain of a songbird: changes in structures associated with mating and song. Behav Brain Res 173:1–10

Janssens E, Dauwe T, Van Duyse E et al (2003) Effects of heavy metal exposure on aggressive behavior in a small territorial songbird. Arch Environ Contam Toxicol 45:0121–0127

Jokimäki J, Suhonen J, Jokimäki-Kaisanlahti ML et al (2014) Effects of urbanization on breeding birds in European towns: impacts of species traits. Urban Ecosyst 19:1565–1577

Kark S, Iwaniuk A, Schalimtzek A et al (2007) Living in the city: can anyone become an 'urban exploiter'? J Biogeogr 34:638–651

Lafleur NE, Rubega MA, Elphick CS (2007) Invasive fruits, novel foods and choice: an investigation of European Starling and American Robin frugivory. Wilson J Ornithol 119:429–438

Laiolo P (2010) The emerging significance of bioacoustics in animal species conservation. Biol Conserv 143:1635–1645

Laiolo P (2011a) The Rufous-collared Sparrow *Zonotrichia capensis* utters higher frequency songs in urban habitats. Rev Catalana Ornitol 27:25–30

Laiolo P (2011b) Homogenisation of birdsong along a natural–urban gradient in Argentina. Ethol Ecol Evol 23:274–287

León E, Beltzer A, Quiroga M (2014) El jilguero dorado (*Sicalis flaveola*) modifica la estructura de sus vocalizaciones para adaptarse a hábitats urbanos. Rev Mex Biodivers 85:546–552

Leveau L (2008) Dynamics of nectarivory in the House Sparrow in a urban environment. Ornitol Neotrop 19:275–281

Levin L, Fajardo L, Ceballos N (2000) Orden de llegada y agresiones en aves urbanas en una fuente de alimento controlada. Ecotropicos 13:75–80

Levis NA, Pfennig DW (2016) Evaluating 'plasticity-first' evolution in nature: key criteria and empirical approaches. Trends Ecol Evol 31:563–574

López-Flores V, MacGregor-Fors I, Schondube J (2009) Artificial nest predation along a Neotropical urban gradient. Landsc Urban Plan 92:90–95

Lopez-Sepulcre A, Kokko H (2012) Behavioural plasticity and environmental change. In: Candolin U, Wong BBM (eds) Behavioural responses to a changing world: mechanisms and consequences. Oxford University Press, Oxford, pp 2–15

Lowry H, Lill A, Wong BB (2013) Behavioural responses of wildlife to urban environments. Biol Rev 88:537–549

MacGregor-Fors I, Blanco-García A, Chávez-Zichinelli C et al (2011) Relación entre la presencia de luz artificial nocturna y la actividad del mosquero cardinal (*Pyrocephalus rubinus*). El Canto del Centzontle 2:64–71

Marzluff JM (2001) Worldwide urbanization and its effects on birds. In: Marzluff J, Bowman R, Donnelly R (eds) Avian ecology and conservation in an urbanizing world. Springer, New York, pp 19–47

Marzluff JM (2016) A decadal review of urban ornithology and a prospectus for the future. Ibis 159:1–13

Maxwell SL, Fuller RA, Brooks TM et al (2016) The ravages of guns, nets and bulldozers. Nature 536:143–145

McKay JL, Maher CR (2012) Relationship between blood mercury levels and components of male song in Nelson's Sparrows (*Ammodramus nelsoni*). Ecotoxicology 21:2391–2397

Mendoça LB, dos Anjos L (2006) Feeding behavior of hummingbirds and perching birds on *Erythrina speciosa* Andrews (Fabaceae) flowers in an urban area, Londrina, Paraná, Brazil. Rev Bras Zool 23:42–49

Miranda AC, Schielzeth H, Sonntag T et al (2013) Urbanization and its effects on personality traits: a result of microevolution or phenotypic plasticity? Glob Chang Biol 19:2634–2644

Moirón M, González-Lagos C, Slabbekoorn H et al (2015) Singing in the city: high song frequencies are no guarantee for urban success in birds. Behav Ecol 26:843–850

Møller AP (2008) Flight distance of urban birds, predation, and selection for urban life. Behav Ecol Sociobiol 63:63–75

Møller AP (2009) Successful city dwellers: a comparative study of the ecological characteristics of urban birds in the Western Palearctic. Oecologia 159:849–858

Møller AP (2010a) The fitness benefit of association with humans: elevated success of birds breeding indoors. Behav Ecol 21:913–918

Møller AP (2010b) Interspecific variation in fear responses predicts urbanization in birds. Behav Ecol 21:365–371

Møller AP (2014) Behavioural and ecological predictors of urbanization. In: Gil D, Brumm H (eds) Avian urban ecology: behavioural and physiological adaptations. Oxford University Press, Oxford, pp 131–142

Møller AP, Erritzøe J (2015) Brain size and urbanization in birds. Avian Res 6:1–7

Møller AP, Grim T, Ibáñez-Álamo JD et al (2013) Change in flight initiation distance between urban and rural habitats following a cold winter. Behav Ecol 24:1211–1217

Müller JC, Partecke J, Hatchwell BJ et al (2013) Candidate gene polymorphisms for behavioural adaptations during urbanization in blackbirds. Mol Ecol 22:3629–3637

Norris KJ (1990) Female choice and the quality of parental care in the Great Tit *Parus major*. Behav Ecol Sociobiol 27:275–281

Norris KJ (1993) Heritable variation in a plumage indicator of viability in male Great Tits *Parus major*. Nature 362:537–539

Ortega-Álvarez R, Calderón-Parra R (2014) Hunting the unexpected: Harris's Hawks (*Parabuteo unicinctus*) preying on bats in a Neotropical megacity. Rev Bras Ornitol 22:297–299

Pacheco-Vargas G, Losada-Prado S (2015) Efecto del ruido del tráfico vehicular en cantos de *Hylophilus flavipes* y *Cyclarhis gujanensis*. Revista Ciencia en Desarrollo 6:117–183

Padilla BJ, Rodewald AD (2015) Avian metapopulation dynamics in a fragmented urbanizing landscape. Urban Ecosyst 18:239–250

Partecke J (2014) Mechanisms of phenotypic responses following colonization of urban areas: from plastic to genetic adaptation. In: Gil D, Brumm H (eds) Avian urban ecology: behavioural and physiological adaptations. Oxford University Press, Oxford, pp 131–142

Pelletier F, Garant D (2012) Population consequences of individual variation in behavior. In: Candolin U, Wong BBM (eds) Behavioural responses to a changing world: mechanisms and consequences. Oxford University Press, Oxford, pp 145–154

Piratelli A, Favoretto GR, de Almeida M (2015) Factors affecting escape distance in birds. Zoologia 32:438–444

Plummer KE, Siriwardena GM, Conway GJ et al (2015) Is supplementary feeding in gardens a driver of evolutionary change in a migratory bird species? Glob Chang Biol 21:4353–4363

Quesada J, MacGregor-Fors I (2010) Avian community responses to the establishment of small garden allotments within a Mediterranean habitat mosaic. Anim Biodivers Conserv 33:53–61

Quesada J, Senar JC (2009) Cross-fostering experiments to compare carotenoid- and melanin-based plumage traits and long-term parental effects in post-moulted Great Tits. Behaviour 146:1245–1251

Rebolo-Ifrán N, Carrete M, Sanz-Aguilar A et al (2015) Links between fear of humans, stress and survival support a non-random distribution of birds among urban and rural habitats. Sci Rep 5:13723

Redondo P, Barrantes G, Sandoval L (2013) Urban noise influences vocalization structure in the House Wren *Troglodytes aedon*. Ibis 155:621–625
Reichard SH, Chalker-Scott L, Buchan S (2001) Interactions among non-native plants and birds. In: Marzluff J, Bowman R, Donnelly R (eds) Avian ecology and conservation in an urbanizing world. Springer, New York, pp 179–223
Ríos-Chelén A, Quirós-Gerrero E, Garcia C (2013) Dealing with urban noise: vermilion flycatchers sing longer songs in noisier territories. Behav Ecol Sociobiol 67:145–152
Rivera López A, MacGregor-Fors I (2016) Urban predation: a case study assessing artificial nest survival in a Neotropical city. Urban Ecosyst 19:649–655
Robertson B, Kriska G, Horvath V et al (2010) Glass buildings as bird feeders: urban birds exploit insects trapped by polarized light pollution. Acta Zool Acad Sci Hung 56:283–293
Rosenthal GG, Stuart-Fox D (2012) Environmental disturbance and animal communication. In: Candolin U, Wong BBM (eds) Behavioural responses to a changing world: mechanisms and consequences. Oxford University Press, Oxford, pp 16–31
Ruiz G, Rosenmann M, Novoa FF et al (2002) Hematological parameters and stress index in Rufous-collared Sparrows dwelling in urban environments. Condor 104:162–166
Sainz-Borgo C (2016) Diet composition of birds associated to an urban forest patch in Northern Venezuela. Interciencia 41:119–126
Salazar A, Garitano-Zavala A (2015) Nest-predator identity as a key factor to understand the differences in nest predation between urban and peri-urban sites in La Paz Metropolitan área. Paper presented on the X Neotropical Ornithological Congress and XXII Congresso Brasileiro de Ornitologia, Manaus, 19–24 July 2015
Salinas-Melgoza A, Salinas-Melgoza V, Wright T (2013) Behavioral plasticity of a threatened parrot in human-modified landscapes. Biol Conserv 159:303–312
San Martín JA (2009) Un ejemplo de explotador urbano: consumo de frutos de árboles exóticos por el jilguero (*Carduelis barbata*), en dos centros urbanos de la zona central de Chile. Nuestras Aves 54:30–32
Senar JC, Conroy MJ, Quesada J et al (2014) Selection based on the size of the black tie of the Great Tit may be reversed in urban habitats. Ecol Evol 4:2625–2632
Seress G, Bókony V, Heszberger J et al (2011) Response to predation risk in urban and rural House Sparrows. Ethology 117:896–907
Shanahan DF, Strohbach MW, Warren PS et al (2014) The challenges of urban living. In: Gil D, Brumm H (eds) Avian urban ecology. Oxford University Press, Oxford, pp 3–20
Shettleworth SJ (2010) Cognition, evolution, and behavior, 2nd edn. Oxford University Press, Oxford
Shochat E, Warren PS, Faeth SH et al (2006) From patterns to emerging processes in mechanistic urban ecology. Trends Ecol Evol 21:186–191
Shochat E, Lerman SB, Anderies JM et al (2010) Invasion, competition, and biodiversity loss in urban ecosystems. Bioscience 60:199–208
Sih A, Ferrari MC, Harris DJ (2011) Evolution and behavioural responses to human-induced rapid environmental change. Evol Appl 4:367–387
Sih A, Trimmer PC, Ehlman SM (2016) A conceptual framework for understanding behavioral responses to HIREC. Curr Opin Behav Sci 12:109–114
Slabbekoorn H, Peet M (2003) Ecology: birds sing at a higher pitch in urban noise. Nature 424:267–267
Snell-Rood EC (2013) An overview of the evolutionary causes and consequences of behavioural plasticity. Anim Behav 85:1004–1011
Sol D, Lapiedra O, González-Lagos C (2013a) Behavioural adjustments for a life in the city. Anim Behav 85:1101–1112
Sol D, González-Lagos C, Moreira D et al (2013b) Measuring tolerance to urbanization for comparative analyses. Ardeola 60:3–13
Sol D, González-Lagos C, Moreira D et al (2014) Urbanisation tolerance and the loss of avian diversity. Ecol Lett 17:942–950

Soto-Gamboa M, Silva-Escobar A, Uribe Y et al (2015) Noise and social environment affects males reproductive calls in *Zonotrichia capensis*: intraspecific variability or plasticity? Paper presented on the X Neotropical Ornithological Congress and XXII Congresso Brasileiro de Ornitologia, Manaus, 19–24 July 2015

Suárez-Rodríguez M, López-Rull I, Garcia CM (2012) Incorporation of cigarette butts into nests reduces nest ectoparasite load in urban birds: new ingredients for an old recipe? Biol Lett 9:20120931

Suárez-Rodríguez M, Montero-Montoya RD, Macías Garcia C (2017) Anthropogenic nest materials may increase breeding costs for urban birds. Front Ecol Evol 5:1–10

Tryjanowski P, Sparks TH, Kuźniak S et al (2013) Bird migration advances more strongly in urban environments. PLoS One 85:e63482

Tuomainen U, Candolin U (2011) Behavioural responses to human-induced environmental change. Biol Rev 86:640–657

United Nations (2015) World urbanization prospects – 2014 revision. United Nations, New York

Vangestel C, Mergeay J, Dawson DA et al (2011) Spatial heterogeneity in genetic relatedness among House Sparrows along an urban–rural gradient as revealed by individual-based analysis. Mol Ecol 20:4643–4653

West-Eberhard MJ (1989) Phenotypic plasticity and the origins of diversity. Annu Rev Ecol Syst 20:249–278

Wingfield JC, Maney DL, Breuner CW et al (1998) Ecological bases of hormone behavior interactions: "The emergency life history stage". Am Zool 38:191–206

Wong BBM, Candolin U (2015) Behavioral responses to changing environments. Behav Ecol 26:665–673

Yeh PJ, Price TD (2004) Adaptive phenotypic plasticity and the successful colonization of a novel environment. Am Nat 164:531–542

Zuria I, Rendón-Hernández G (2010) Notes on the breeding biology of common resident birds in an urbanized area of Hidalgo, Mexico. Huitzil 11:35–41

Chapter 7
Warning! Urban Threats for Birds in Latin America

Diego Santiago-Alarcon and Carlos A. Delgado-V

Abstract Wild birds are subject to a diverse array of natural causes of mortality, such as predation and parasitism. However, anthropogenic sources are becoming major threats for birds, particularly in urban systems, where major ecological impacts can affect both intra- and interspecific interactions. For instance, host-parasite interactions are modified in such a way that parasites can start appearing in novel hosts and generate health problems. Furthermore, nonnative predators can severely affect bird populations, at times driving them locally extinct when occurring in combination with the reduction of suitable sites. In this chapter, we show that urbanization entails such drastic alterations on the environment that antagonistic interactions can become an important threat to birds. Predation by cats has been identified as the most important threat to urban birds, accounting for up to billions of deaths annually in the USA alone, followed by bird collisions with building structures. Our review reveals a lack of knowledge related to the main urban bird threats in Latin America. The available information suggests that cats and collisions with building structures and vehicles might also be major sources of bird mortality in the region. However, it is premature to make generalizations at this stage because Latin American cities develop differently and are immersed in diverse socioecological contexts (i.e., different cultural habits imposed by local environmental conditions). We suggest that systematic studies in urban Latin America should focus on three main areas: (i) predation by cats; (ii) building and vehicle collisions; and (iii) alterations of host-parasite interactions.

Keywords Avian parasitology • Building collisions • Host-parasite interactions • Urban ecology • Urban predators

D. Santiago-Alarcon (✉)
Red de Biología y Conservación de Vertebrados, Instituto de Ecología, A.C. (INECOL), Carretera antigua a Coatepec 351, El Haya, Xalapa 91070, Mexico
e-mail: diego.santiago@inecol.mx

C.A. Delgado-V
Facultad de Ciencias y Biotecnología, Universidad CES,
Calle 10 A No. 22-04, Medellín, Colombia

I. MacGregor-Fors, J.F. Escobar-Ibáñez (eds.), *Avian Ecology in Latin American Cityscapes*, DOI 10.1007/978-3-319-63475-3_7

7.1 Global Perspective

In wild bird populations, mortality can result from regular ecological dynamics, such as density-dependent events that determine the size of a population in a given locality (e.g., population carrying capacity; Stevens 2009). Furthermore, mortality also occurs from interspecific interactions such as parasitism (Palinauskas et al. 2008), as well as from direct and indirect anthropogenic causes (e.g., pollution, habitat destruction, domestic animal predation, collisions; Erickson et al. 2005; Loss et al. 2015). These anthropogenic factors are becoming the most worrisome threats around the globe because they translate into billions of bird deaths annually. However, apart from the USA and Canada in North America, the UK in Europe, and Australia and New Zealand in Oceania, information on bird mortality from anthropogenic causes is scarce or inexistent (Loss et al. 2015). Additionally, there is a similar lack of parasitological studies of wild birds in anthropogenically modified systems (Delgado-V and French 2012; Hernández-Lara et al. 2017).

The main anthropogenic causes of wild bird mortality in urban areas of the USA and Canada, where most information is available, are predation by cats, collisions with building windows, and vehicles (Loss et al. 2015; Table 7.1); yet, the dearth of knowledge of avian parasitism in urban areas does not allow us to gauge the impact of altering host-parasite interactions by anthropogenic factors on bird health. Cats have decimated bird populations on small islands (Nogales et al. 2013) and are estimated to kill, based on statistical extrapolations, between one and four billion wild birds annually in the USA alone (Loss et al. 2013). In Canada, the estimate ranges from 100 to 350 million birds killed by cats (Blancher 2013), of which ~186 million are breeding individuals (Calvert et al. 2013). There is little information on the numbers of birds killed by cats in terms of bird species or avian guilds, and on the negative impacts that such a factor has on wild bird populations, particularly those of

Table 7.1 Estimates of bird mortality from anthropogenic urban factors in the USA and Canada

Mortality source[a]	Estimate	95% CI	Country
Cat predation	2,407,000,000	1,306,000,000–3,992,000,000	USA
	204,000,000	105,000,000–348,000,000	Canada
Building collisions	599,000,000	365,000,000–988,000,000	USA
	24,900,000	16,100,000–42,200,000	Canada
Vehicle collisions	199,600,000	88,700,000–339,800,000	USA
	13,810,906	8,914,341–18,707,470	Canada
Power line collisions	22,800,000	7,700,000–57,300,000	USA
	25,600,000	10,100,000–41,200,000	Canada
Power line electrocutions	5,630,000	920,000–11,550,000	USA
	481,399	160,836–801,962	Canada

Data source: Loss et al. (2015)

[a]Estimates are mortalities for individual birds; they do not include destroyed nests, eggs, and nestlings

conservation concern (e.g., Loyd et al. 2013). In Canada, land birds are the group suffering higher impacts, mainly species foraging on the ground, using winter feeders, and nesting on or close to the ground (Blancher 2013; Calvert et al. 2013). The reduction of cat predation on birds has been suggested to be easily implemented for pets, via environmental education programs, but for feral cats, control can be controversial when it implies lethal control due to its social implications (McCarthy et al. 2012; Loss et al. 2015).

Bird mortality by collision with building structures (mostly windows) in the USA and Canada ranges from 16 million to close to a billion individuals (Machtans et al. 2013; Loss et al. 2014a). Long-distance migrant species, some of conservation concern, are the most vulnerable group to this threat, particularly during the migration period that contrasts with the low mortality rates recorded in urban-dwelling species (Arnold and Zink 2011). Although limited, some strategies have already been implemented in the USA to reduce window collisions, for example, by using UV reflecting surfaces (Klem and Saenger 2013), and by placing patterns and objects on the exterior surface to reduce reflective window area (Klem et al. 2009). Others suggest that such methods serve as attractants, instead of deterring visitation by birds; however, they argue that due to this attraction, birds would be aware of windows reducing collision likelihood (Habberfield and Clair 2016). Another major type of avian collision is that with vehicles, which ranges from 13 to more than 300 million deaths per year in the USA and Canada (Bishop and Brogan 2013; Loss et al. 2014b). Scarce information exists regarding avian groups at greater risk, how mortality varies as a function of road features, and there are woefully no tested strategies to reduce this mortality risk (Loss et al. 2015).

Other sources of mortality include the collision with communication towers, wind turbines, and power lines (electrocution included), which generate bird deceases in the range of 200,000 to more than 50 million altogether (Longcore et al. 2012; Loss et al. 2013, 2015; Smallwood 2013). Some of these estimates are reliable given that they took into account sources of bias such as scavenging rates, search efficiency, and sampling effort (Longcore et al. 2012). When focusing on collisions with communication towers, it has been reported that the annual mortality from this source might represent >1% of their entire populations in some species, of which apparently migratory songbirds are at greater risk (Loss et al. 2015). The factors that seem to be most relevant for increased risk of bird mortality are tower height, turbine placement, power lines materials and placement, and signaling red lights at towers (Manville 2009; Loss et al. 2015). Bird mortality seems to be low at wind facilities compared to other sources, but the current and rapid expansion of wind energy might substantially increase current estimates up to a fourfold in the USA within the next 20 years (Loss et al. 2015). Hence, it is of the highest relevance to continue, improve, and expand bird surveys in relation to the design and construction of wind farms.

Different to the other mentioned major threats, parasitism is a natural process that has shown to be one of the most successful ways of life on Earth, which has independently evolved in animals at least 60 times (Poulin and Morand 2000; Poulin 2007). However, parasites do not necessarily spell disaster (as often perceived), they actually are essential for healthy, functional ecosystems (Marcogliese 2005; Lafferty et al. 2008). Major problems with parasites result when they become pathogens (i.e., parasites that generate health problems in their hosts), which can be due to a weakened immune system or environmental changes (e.g., Hamer et al. 2012). Currently, the main parasitological challenges from the medical and veterinary trenches are the rapid modifications ensued by anthropogenic activities, which are behind global viral outbreaks (Jones et al. 2008; Aguirre et al. 2012). For example, biodiversity loss opens opportunities for novel host-parasite associations, which can result in epidemics (e.g., West Nile virus, amphibian viruses; Kilpatrick et al. 2006; Johnson et al. 2013).

Urbanization, considered a major threat to biodiversity, is known to alter the ecology of wildlife pathogens (Bradley and Altizer 2007; Shochat et al. 2010). By modifying among-species contact rates, urbanization changes parasite abundance and distribution by altering vector-feeding preferences (e.g., avian malaria; Santiago-Alarcon et al. 2012, 2013). Despite their relevance, parasitological studies in urbanized areas are scarce and far behind projects involving macroorganisms such as birds and mammals (e.g., MacGregor-Fors et al. 2015), limiting our capacity to understand the ecological patterns and processes in urban ecosystems from the parasitological standpoint (Delgado-V and French 2012; Martin and Boruta 2014; Santiago-Alarcon et al. 2016; Carbó-Ramírez et al. 2017). For instance, for vector-borne avian parasites, the information is limited to lists of pathogens and some of their effects on urban-dwelling birds (e.g., Rock Pigeon – *Columba livia*, House Sparrow – *Passer domesticus*; Delgado-V and French 2012). In addition, there is no common pattern in infection rates when comparing urban and nonurban conditions at different geographical locations, either for the same or for different bird species; some bird species have higher infection rates in urban compared to nonurban counterparts, and the opposite pattern can be observed for the same species at other localities (Fokidis et al. 2008; Evans et al. 2009; Hernández-Lara et al. 2017; Table 7.2). Urbanization effects are complex and might be site-, parasite-, and host-specific (Martin and Boruta 2014; Delgado-V and French 2015). Thus, parasitological studies in urban systems with proper controls (i.e., nonurban well-preserved regional system) will clarify the relationship between urbanization and host-parasite ecological dynamics (Hernández-Lara et al. 2017).

Despite current efforts, the impact of the above mentioned factors on urban bird populations need to be better gauge, particularly for specific groups of birds (e.g., Nearctic-Neotropical migrants, geographically restricted species; Erickson et al. 2005). Reasons include nonstandardized sampling methods, observation biases, and scavenging rates, all of which vary as a function of avian group, ecosystem, and seasonality (Erickson et al. 2005; Loss et al. 2015). Hence, the implementation of random and representative number of experimental units along with standardized sampling methods is necessary to define the real impact of the factors mentioned above on birds. Furthermore, detection and removal (scavenging) rates must be estimated at each study site because local conditions will certainly change the quantification of mortalities, which are necessary to adjust impact estimates (Erickson et al. 2005).

Table 7.2 Prevalence (i.e., proportion of infected birds in a sample) estimates from studies on avian parasites conducted in different parts of the world that include at least one urban and nonurban comparison

	Blood parasite prevalence (%)		
City (country)	Urban	Nonurban	Source[a]
Munich (Germany) – Spring	58	73	Geue and Partecke (2008)
Munich (Germany) – Summer	75	100	Geue and Partecke (2008)
Phoenix (USA)	0	44	Fokidis et al. (2008)
Phoenix (USA)	22	0	Fokidis et al. (2008)
Phoenix (USA)	23	39	Fokidis et al. (2008)
Phoenix (USA)	0	0	Fokidis et al. (2008)
Berlin (Germany)	62	95	Evans et al. (2009)
Groningen (Germany)	90	75	Evans et al. (2009)
Krakow (Poland)	78	75	Evans et al. (2009)
Madrid (Spain)	25	78	Evans et al. (2009)
Valencia (Spain)	38	80	Evans et al. (2009)
Prague (Czech Republic)	82	90	Evans et al. (2009)
Riga (Latvia)	62	63	Evans et al. (2009)
Sheffield (England)	43	83	Evans et al. (2009)
Tallinn (Estonia)	58	39	Evans et al. (2009)
Tunis (Tunisia)	80	98	Evans et al. (2009)
Xalapa (México)	63	38	Hernández-Lara et al. (2017)

[a]Prevalence values are approximate rounded figures acquired from cited sources

7.2 Latin American Studies

In this section, we review studies related to the potential threats imposed by urbanization to birds in Latin American cities. For this, we searched for publications in the Web of Science (www.webofknowledge.com) and Scopus (www.scopus.com) platforms using the following keywords in both English and Spanish: 'bird,' 'avian,' 'urban,' 'Latin America.' We also searched for theses and dissertations in databases from major Latin American universities (e.g., National Autonomous University of Mexico – UNAM, University of São Paulo – USP, National University of Colombia – UNAL). Finally, we consulted urban ecology and urban threat-related expert colleagues from Latin America for further additional publications that could enrich our review.

7.2.1 Exotic, Invasive, and Domestic Predators

With human populations increasing worldwide along with the lack of proper regulatory measures to own pets in most Latin American countries, domestic cat populations are likely growing in parallel, a situation that might bring associated risks to local wildlife (e.g., van Heezik et al. 2010). Although there is no available published information that documents the effects of cats on urban wildlife in Latin America, there are initial efforts to gather information, such as Aburrá Natural (www.aburranatural.org), which has been collecting data about cat predation in Medellín, Colombia. Preliminary results show that cat predation is more prevalent at the interior of the urban area (68%, $n = 25$ kills, average kills = 0.002/ha) when compared to its periphery and natural areas ($n = 11$ kills, average kills = 0.0004/ha). At least 20 bird species have been recorded to be depredated by cats, including young individuals of the endemic Colombian Chachalaca (*Ortalis columbiana*) and the near-endemic Scrub Tanager (*Tangara vitriolina*).

In Mexico, we found a study analyzing the diet of both domestic and feral cats in an urbanization intensity gradient of the city of Morelia. Small mammals and reptiles comprised most of the studied cat diet in the urban and nonurban environments; only in pine-oak forests, birds became dominant. When cat scats were analyzed, small mammals remained as the main item of cat diet in the urban environment, with birds and reptiles representing a minor proportion (Orduña-Villaseñor 2015). Results from this study suggest that cats prefer mammals instead of birds in Morelia, which might be unexpected given information from other countries (e.g., Loss et al. 2015). Based on this study, we suggest that a larger array of potential preys, given by the biodiversity present in tropical latitudes, could explain the contrasting pattern to what is known from temperate regions.

Regarding dogs, they might also be implicated as a factor threatening birds in urbanized landscapes. For instance, the occurrence of dogs might be one of the causes of unsuccessful nesting of the Spot-flanked Gallinule (*Porphyriops melanops*) in a channel of pluvial drainages connected to urban wetlands (Becerra-Galindo et al. 2005). In Mexico, a study conducted in three cities of the state of Michoacán showed that dog numbers are negatively associated with the diversity of moderately abundant and rare bird species, suggesting that feral and off-leash dogs are stressful agents and potential bird predators in cities (MacGregor-Fors and Schondube 2011). Additionally, there is a growing evidence of records of free-ranging domestic dogs within and around South American cities, for instance, in Medellín and Bogota; however, their local effects on birds and other wildlife remain unstudied (Delgado-V pers. obs.).

7.2.2 Collision with Building Structures

Information regarding collisions with building windows is receiving a growing, yet still scarce, scientific attention in Latin America (Cupul-Magaña 2003; Agudelo-Álvarez 2006; Agudelo-Álvarez et al. 2010; Menacho-Odio 2015; Oviedo 2014). Different species, sizes, and feeding guilds, from hummingbirds to chachalacas (e.g., Colombian Chachalaca; Fig. 7.1; Delgado-V pers. obs.), have been recorded crashing on windows. In particular, many collisions correspond to Nearctic-Neotropical migrants, such as the Yellow-breasted Chat (*Icteria virens*),

Fig. 7.1 Fatal collision of a Colombian Chachalaca (*Ortalis columbiana*) on a building located at Universidad CES, Medellín, Colombia (Photo: Hana Londoño)

Ruby-throated Hummingbird (*Archilochus colubris*), as well as Indigo Bunting (*Passerina cyanea*; M. A. Gómez-Martínez per. comm.; Fig. 7.2) in Xalapa (Mexico), and Swainson's Thrush (*Catharus ustulatus*), Northern Waterthrush (*Parkesia noveboracensis*), and Mourning Warbler (*Geothlypis philadephia*) in Medellín (Colombia). In Bogota, Colombia, approximately 88% of recorded collisions are fatal, and the most frequent group to collide are also Nearctic-Neotropical migrants (Agudelo-Álvarez 2006), such as the Red-eyed Vireo (*Vireo olivaceus*), Swainson's Thrush, Scarlet Tanager (*Piranga olivacea*), and Summer Tanager (*Piranga rubra*). Window collisions have been recorded to be more frequent in buildings located near urban greenspaces in Mexico and Colombia (Cupul-Magaña 2003; Agudelo-Álvarez et al. 2010; Delgado-V and Correa-H. 2013). According to Ocampo-Peñuela et al. (2016), decals can prevent bird-window collisions at small nonurban Colombian residences, finding a reduction of 84% of strikes after bird decal application in ~36 m^2 of total glass area. This reduction in window collision rate might be due to the awareness of birds given certain cues on windows (e.g., decals, UV light), as suggested by Habberfield and Clair (2016).

Collisions with other type of infrastructure in urban and peri-urban areas, such as electric power lines or communication towers, have been poorly documented. However, similar to collisions with buildings, migratory birds are the most affected

Fig. 7.2 Fatal collision of a Nearctic-Neotropical migrant passerine (Indigo Bunting – *Passerina cyanea*) on a three-story building located within the city of Xalapa with important reflecting pane surface (Photo: Miguel A. Gómez-Martínez)

group by such urban structures. Álvarez-López (1993) reported that at least nine migratory species collided with towers: the Swainson's Thrush had about 76% of fatal collisions, while the remaining 24% of collisions were accounted for the Blackburnian Warbler (*Setophaga fusca*), Canada Warbler (*Cardellina canadensis*), Black-and-white Warbler (*Mniotilta varia*), Bay-breasted Warbler (*Setophaga castanea*), Northern Waterthrush, and Mourning Warbler.

7.2.3 Collisions with Vehicles

The ecology of roads in urban areas and urbanized landscapes has been poorly studied (van der Ree et al. 2009) and Latin America is not an exception. Although Brazil has history on road ecology, research of this type has been mostly conducted in nonurban landscapes (e.g., Barri 2010; Bager and da Rosa 2012; da Rosa and Bager 2012; De La Ossa-Nadjar and De La Ossa 2015). A study performed in areas surrounding Andean cities showed that mammals are the most affected vertebrate group regarding collisions with vehicles (Delgado-V 2007, 2014). However, the diversity of birds found dead by car collisions within and around Medellín

(Colombia) is growing, as suggested by previous and unpublished results (Delgado-V pers. obs.), where some endemics and vulnerable species (sensu Bird Life International 2016) have been recorded, such as the Red-bellied Grackle (*Hypopyrrhus pyrohypogaster*).

7.2.4 Parasites

Although still few, studies focused on bird parasites in urban Latin America are growing. In Brazil, Belo et al. (2011) found higher prevalence (i.e., proportion of infected birds in a sample or population sample) of parasites from the genera *Plasmodium* and *Haemoproteus* in an urban site when compared to nonurban areas. A study conducted in a Brazilian urbanization intensity gradient showed that there is a negative relationship between bird parasite species richness and the degree of urbanization (Calegaro-Marques and Amato 2014), indicating that urbanization has a negative effect on the number of parasite species, which is similar to patterns reported for other organisms like fishes and plants (Alberti 2008). However, there was a positive association between environmental heterogeneity and parasite richness, suggesting that even when urbanization can disrupt host-parasite interactions, some systems can actually benefit from the heterogeneity provided by cities (Calegaro-Marques and Amato 2014).

In Mexico, Hernández-Lara et al. (2017) found higher prevalence of *Plasmodium* parasites (Fig. 7.3) infecting Chestnut-capped Brushfinches (*Arremon brunneinucha*) in an urban greenspace compared to a peri-urban forest, a well-preserved cloud forest, a shade coffee plantation, and a cattle ranch. Carbó-Ramírez et al. (2017) compared three environmentally contrasting urban greenspaces (two in Mexico and one in Germany), finding that despite different environmental conditions and avian composition, haemosporidian assemblages had similar abundance distribution, similar infection rates, and phylogenetically diverse parasite communities (i.e., genetically different parasites) dominated by widespread generalist lineages. Finally, a comparison of urban and nonurban populations of the nonnative House Sparrow in Mexico City shows that there is a lower risk of avian malaria infections in the city, where birds tend to use highly developed areas and avoid greenspaces (Santiago-Alarcon et al. in prep.). Thus, this nonnative urban exploiter seems to have been released from its natural parasites in highly developed areas within cities. In fact, some House Sparrow populations from Brazil showed lower haemosporidian prevalence compared to native birds and also to European populations (Lima et al. 2010; Marzal et al. 2011).

In Brazil, a study reported a 22.3% prevalence of *Toxoplasma gondii* from Eared Doves (*Zenaida auriculata*) sampled in areas with human influence (i.e., cropland, farm, university campus), and another study, also developed in Brazil, reported 25% infections in the House Sparrow for the same parasite (Gondim et al. 2010). Such results are relevant because high rates, as those reported above, may develop into human and veterinary health problems. *T. gondii* is highly prevalent in humans

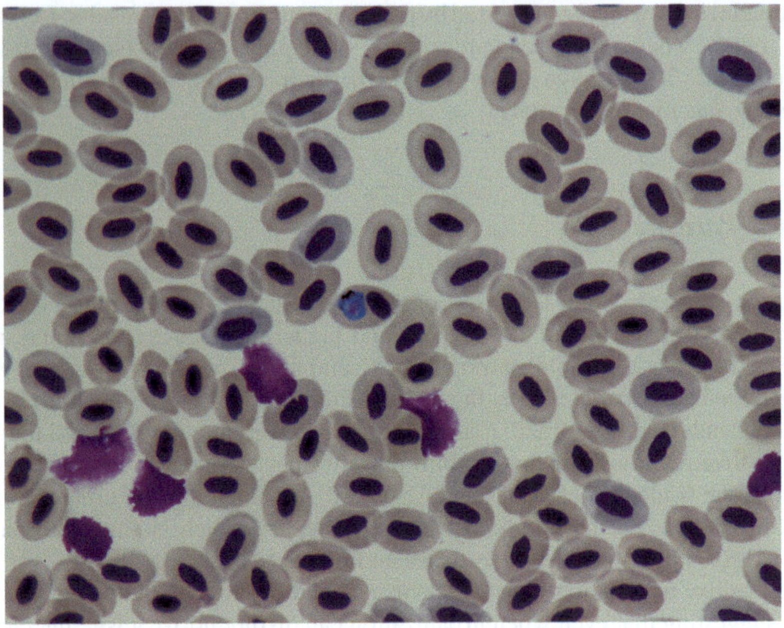

Fig. 7.3 A parasite of the genus *Plasmodium* subgenus *Haemamoeba* (a macrogametocyte), infecting a Chestnut-capped Brushfinch (*Arremon brunneinucha*) captured in an urban forest. The parasite is observed in the center, it displaces the nucleus of the erythrocyte towards the right side; all other surrounding erythrocytes are uninfected. Some immature erythrocytes are observed (i.e., those with a bluish-violet cytoplasm and an enlarged nucleus compared to mature erythrocytes) (Photo: DS-A)

around the world, but clinical cases (i.e., symptoms and disease) are less common (Roberts and Janovy 2009). In particular, *T. gondii* is a common parasite in domestic cats, which can be transmitted to humans and generate illness in pregnant woman (i.e., toxoplasmosis), producing birth defects, abortions, and even the death of fetuses (Roberts and Janovy 2009). Acute infections produce symptoms such as pain and swollen lymph nodes in the cervical and supraclavicular region, along with fever, headache, muscle pain, anemia, and sometimes lung problems (Roberts and Janovy 2009). In fact, a worrisome aspect in urban areas is that nonnative abundant bird species, such as doves and sparrows, can serve as reservoirs for parasites of health relevance (e.g., *T. gondii*, *Neospora caninum*; de Sousa et al. 2010, Gondim et al. 2010). Other parasites of veterinary importance detected in doves and pigeons sampled in Puerto Ayora (Galápagos) include *Trichomonas gallinae* with 44% prevalence (Padilla et al. 2004). *Salmonella* spp. have been detected infecting Rock Pigeons (8% prevalence) in São Paulo (Brazil), and in the same study, some pigeons were positive to serological tests against Newcastle disease virus (de Sousa et al. 2010). Similarly, a report from the state of Rio Grande do Sul (Brazil) has found prevalence rates ranging from 2.9 to 26.5% for different species of nematodes, infecting the Picui Ground-Dove (*Columbina picui*; Coimbra et al. 2009). Finally,

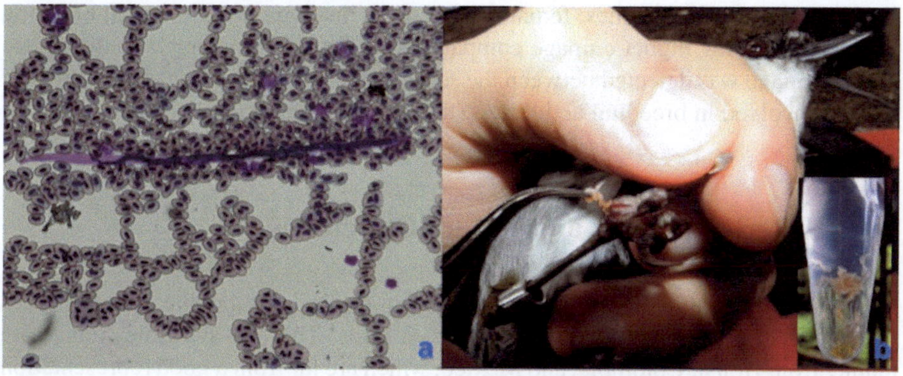

Fig. 7.4 A microfilaria or larva from a roundworm, surrounded by bird erythrocytes (**a**). Adult round worms of the genus *Pelecitus*, extracted from a foot nodule of a Chestnut-capped Brushfinch, a plastic tube shows all nematodes extracted from foot nodules (**b**) (Photos: DS-A)

infecting the Chestnut-capped Brushfinch, nematodes from the genus *Pelecitus* were found at both urban and peri-urban greenspaces in Xalapa (Mexico) (both larval stages in the blood and adult worms forming nodules on birds' legs; Santiago-Alarcon pers. obs.; Fig. 7.4).

In terms of viral diseases, which are sometimes of higher public concern due to their rapid adaptation to new hosts, researchers from Puerto Rico found that both native (Greater Antillean Grackle – *Quiscalus niger*) and nonnative (House Sparrow) urban exploiters were readily exposed to West Nile virus, particularly young birds of less than one-year-old (Komar et al. 2012). In Guatemala, this virus was isolated from sentinel domestic chickens at both urban and nonurban locations, but its main competent reservoir host was the Great-tailed Grackle (*Quiscalus mexicanus*). The authors of this study suggest that these grackles are the amplifying host in their study sites (Morales-Betoulle et al. 2013). Higher seroconversion (i.e., positive immunological tests, indicating exposure to the pathogen) of chickens was correlated with high temperature and low rainfall, both at urban and nonurban sites. The virus was also isolated from mosquitoes (i.e., *Culex quinquefasciatus*), which are considered a competent West Nile virus vector at other locations (Morales-Betoulle et al. 2013). Thus, current parasitological studies in Latin America suggest that native and nonnative birds using urban environments are susceptible to parasites with medical and veterinary importance.

Few studies have been focused on bird ectoparasites in urban Latin America. One interesting exception is a study in Mexico City, where cigarette butts found in nests of House Sparrows and House Finches (*Haemorhous mexicanus*) acted as an ectoparasite repellent (Suárez-Rodríguez et al. 2012). This study detected a reduction of nest-dwelling ectoparasites on nests built with cigarette butts. However, recent results show that this benefit could be counterbalanced by chemical damage on red blood cells of such birds more involved in nest building and incubation: an effect that is directly related to the amount of cigarette butts used in nest building.

In the case of House Finches, females were more involved in nesting and breeding activities, showing higher damage compared to males. Contrastingly, no difference in cell damage was detected between the sexes of House Sparrows, as both sexes are equally involved in breeding activities (Suárez-Rodríguez et al. 2017).

7.3 Knowledge Gaps and Future Directions

Taking into account all of the above, it is evident that there is a major lack of studies in urban Latin America focused on threats to birds. The few available studies are mainly from Brazil, Colombia, and Mexico, which are far from a good representation of the diversity of urban scenarios and surrounding environments across Latin America. This is worrisome because there is no current way in which we can measure the impact of such urban factors on Latin American bird communities and their emergent properties, which could be drivers of the general patterns identified for birds in the region (Ortega-Álvarez and MacGregor-Fors 2011; Chaps. 2, 3, 4, and 5). It is also clear that the major bird mortality source in cities is cat predation, either from feral or domestic cats (Loss et al. 2015). The other major sources of bird mortality in urban centers are buildings (particularly bird-window collisions), and collisions with vehicles and power lines. Finally, there is still a long way to know the effect of parasitism on bird health in the region, and we only glimpse at how ecological dynamics of host-parasite interactions are changing due to the urbanization process (e.g., Hernández-Lara et al. 2017). Hence, we propose the following research agenda focused on the major identified and potential threats to birds in urban Latin America.

7.3.1 Cat Predation

Systematic and informal information is needed to understand the nature and impacts of cat predation in urban Latin America; for example, questionnaires and data sheet reports, where cat owners can fill out the number of birds that their cats have retrieved and to have a historical estimate of cat predation based on owner memory. Also, placing cameras and trackers on cats in order to have more reliable measures of predation within and in the periphery of cities could widen our understanding of the hazard that cats represent for urban birds. This last procedure, in combination with systematic bird censuses through citizen science projects and intensive field searches, can be used to weigh estimates based on population densities of different bird species and guilds across time.

7.3.2 *Bird Collisions with Urban Structures*

As suggested above for cat predation, both systematic and informal approaches should be followed to maximize the amount of information. In particular, we suggest two steps to fill this gap: (i) conducting systematic surveys across cities periodically (e.g., monthly), taking into account factors such as road size, closeness to greenspaces, water sources, traffic rate, building height, number and dimension of windows, and density of power lines per square kilometer; (ii) once the main factors associated with this threat are identified, the use of cameras to continuously monitor such structures could aid in the improvement of collision estimates, as there are birds that strike but do not die from the impact. The suggested scheme could shed important light on how mortality varies seasonally and which are the most affected bird groups. One important thing to consider when estimating the effect of this type of threat is carcass decomposition and removal rates, as the information of dead birds will be lost in places where these rates are higher. Thus, knowing such rates will improve mortality estimates and avoid underestimations (see Hager et al. 2012; Bishop and Brogan 2013; Blancher 2013).

7.3.3 *Parasitism*

Performing systematic bird parasitism studies having urban and nonurban comparisons as a minimum for each study area, preferably including an urbanization intensity gradient. It is also important to conduct studies within cities because urban environments are highly heterogeneous. For instance, the density of hard relatively impervious structures might affect bird breeding biology, increasing bird density and the likelihood of disease transmission, also food provisioning via bird feeders can increase contact rates between healthy and diseased animals (Martin and Boruta 2014). Although we are starting to have a clearer idea of the ecology of parasites in Latin American urban systems (e.g., Medellín, Colombia; Mexico City and Xalapa, Mexico; Porto Alegre, Brazil), we are still far from knowing both the real impact on bird health and how host-parasite interactions are modified with urbanization. Certainly, the way in which bird diversity is modified by urbanization will have a direct impact on parasite ecological dynamics (e.g., dilution effect against amplification effect). Hence, answers to basic questions in order to manage veterinary and zoonotic threats are urgently needed, for instance: (i) which host species is a parasite of health importance infecting? (ii) how have parasite preferences changed from nonurban to urban settings? (iii) which hosts are super spreaders and which ones are dilution hosts?

7.3.4 *Airports*

Although not necessarily and directly related to urban threats, airplane strikes with birds are a relevant study topic. Although published studies are scarce in the region, reports from the USA show an increase in the number of strikes between 1990 (1851 strikes) and 2013 (11,399 strikes; Dolbeer 2015). Most importantly, there has been an increase in the number of documented species involved in aircraft strikes (Dolbeer 2015), which can certainly aid in the development of prevention plans. In a study conducted in Brazil (Parnaíba International Airport), several thousand bird individuals from different species have been recorded either directly landing on the runway or flying through landing or take-off routes (Oliveira Cardoso et al. 2014). It is important to consider that airports, even within large urban areas, often have vegetation and water sources that tend to concentrate wildlife species (Patricia Ramírez-Bastida pers. comm.). Bird mortality records from airports are present in unpublished technical reports throughout Latin America; thus, in addition to more studies, we urge airports and related governmental institutions to make the information regarding their bird strike studies available, so a better assessment of airports as bird threats can be conducted, as well as management plans can be developed.

7.3.5 *Rodenticides*

Rat poisons are typically nonspecific rodent control chemicals for rats and mice. Although they have been commonly used to control the Black Rat (*Rattus rattus*), Norway Rat (*Rattus norvegicus*), and House Mouse (*Mus musculus*), anticoagulant poisons are also toxic for other vertebrates. In Latin America, the effects of anticoagulants in wildlife have not been studied in detail, but some evidence suggests that nocturnal and diurnal raptors are a group of special vulnerability due to secondary and tertiary poison consuming via the food chain. In Colombian cities, for example, owls frequently prey on introduced rodents (Delgado-V et al. 2005), some of which have been tested positive for anticoagulants in at least two cities (i.e., Medellín and Bogota, Colombia). Hence, systematic toxicology studies are urgently needed in Latin American cities to know the magnitude of this threat and mitigate its ecological effects.

Once information related to the major threats for birds in urban Latin America becomes available, we will be able to develop evidence-based management and conservation strategies directed towards relevant factors reviewed in depth above. It is difficult to predict which factors will be the most relevant for different cities in Latin America, most likely they will not be the same, even when compared to better known regions (e.g., USA, Canada, Western Europe, and Australia). This is because each city is a unique system with a particular history and development. Although we expect that some general patterns will arise due to the urbanization process, such as avian parasite communities dominated by generalist species, higher infection rates at places with food provisioning (e.g., feeders), higher predation on bird species that inhabit and breed close to the ground, and increased collision rates for buildings that are taller and have a larger surface covered by reflecting windows, it is certainly premature to make generalizations at this stage.

Acknowledgments The authors thank Patricia Ramírez Bastida for comments and suggestions that enhanced the clarity of our manuscript. DS-A was supported by a CONACYT-SEP (CB-2011-01-168524) grant.

References

Agudelo-Álvarez L (2006) Colisión de aves contra los ventanales del campus de la Universidad Javeriana, sede Bogotá. Alternativas de mitigación. Ecología de Aves. Universidad Javeriana, Facultad de Estudios Ambientales y Rurales

Agudelo-Álvarez L, Moreno-Velásquez J, Ocampo-Peñuela N (2010) Colisiones de aves contra ventanales en un campus universitario de Bogotá, Colombia. Ornitol Colomb 10:3–10

Aguirre AA, Ostfeld RS, Daszak P (2012) New directions in conservation medicine: applied cases of ecological health. Oxford Univ Press, New York

Alberti M (2008) Advances in urban ecology: integrating humans and ecological processes in urban ecosystems. Springer, New York

Álvarez-López H (1993) Mortandad de aves migratorias por colisión con una torre de radio. Boletín SAO IV:5–6

Arnold TW, Zink RM (2011) Collision mortality has no discernible effect on population trends of north American birds. PLoS One 6:e24708

Bager A, da Rosa CA (2012) Impacts of the BR-392 highway on bird communities in extreme southern Brazil. Rev Bras Ornitol 20:30–39

Barri FR (2010) Evaluación preliminar de la mortandad de mastofauna nativa por colisión con vehículos en tres rutas Argentinas. Ecología Aplicada 9:161–165

Becerra-Galindo LF, Benítez-Castañeda HD, Cely-Fajardo JE et al (2005) Notas sobre la anidación no exitosa de la Tingua moteada (*Gallinula melanops*) en un canal artificial del humedal Jaboque, Bogotá. Boletín SAO XV:29–38

Belo NO, Pinheiro RT, Reis ES et al (2011) Prevalence and lineage diversity of avian haemosporidians from three distinct cerrado habitats in Brazil. PLoS One 6:e17654

Bird Life International (2016) Hypopyrrhus pyrohypogaster. The IUCN Red List of Threatened Species. http://dx.doi.org/10.2305/IUCN.UK.2016-3.RLTS.T22724268A94857024.en. Accessed 17 Mar 2017

Bishop CA, Brogan JM (2013) Estimates of avian mortality attributed to vehicle collisions in Canada. Avian Conserv Ecol 8:2

Blancher P (2013) Estimated number of birds killed by house cats (*Felis catus*) in Canada. Avian Conserv Ecol 8:3

Bradley CA, Altizer S (2007) Urbanization and the ecology of wildlife diseases. Trends Ecol Evol 22:95–102

Calegaro-Marques C, Amato SB (2014) Urbanization breaks up host-parasite interactions: a case study on parasite community ecology of Rufous-bellied Thrushes (*Turdus rufiventris*) along a rural-urban gradient. PLoS One 9:e103144

Calvert AM, Bishop CA, Elliot RD et al (2013) A synthesis of human-related avian mortality in Canada. Avian Conserv Ecol 8:11

Carbó-Ramírez P, Zuria I, Schaefer HM et al (2017) Avian haemosporidians at three environmentally contrasting urban greensapaces. J Urban Ecol 3:juw011
Coimbra MAA, Mascarenhas CS, Krüger C et al (2009) Helminths parasitizing *Columbina picui* (Columbiformes: Columbidae) in Brazil. J Parasitol 95:1011–1012
Cupul-Magaña FB (2003) Nota sobre colisiones de aves en las ventanas de edificios universitarios en Puerto Vallarta, México. Huitzil 4:17–21
da Rosa CA, Bager A (2012) Seasonality and habitat types affect roadkill of neotropical birds. J Environ Manag 30:1–5
de Sousa E, Berchieri AJ, Pinto AA et al (2010) Prevalence of *Salmonella* spp. antibodies to *Toxoplasma gondii*, and Newcastle disease virus in feral pigeons (*Columba livia*) in the city of Jaboticabal, Brazil. J Zoo Wildl Med 41:603–607
De La Ossa-Nadjar O, De La Ossa JV (2015) Vehicle collisions with wild fauna on the two roads that pass through the montes de María, Sucre, Colombia. Rev U D C A Act & Div Cient 18:503–511
Delgado-V CA (2007) Muerte de mamíferos por vehículos en la vía del Escobero, Envigado (Antioquia), Colombia. Actual Biol 29:235–239
Delgado-V CA (2014) Adiciones al atropellamiento vehicular de mamíferos en la vía de El Escobero, Envigado (Antioquia), Colombia. Revista EIA 11:145–151
Delgado-V CA, Correa-H JC (2013) Estudios ornitológicos urbanos en Colombia: Revisión de literatura. Ing Cienc 19:216–237
Delgado-V CA, French K (2012) Parasite-bird interactions in urban areas: current evidence and emerging questions. Landsc Urban Plan 105:5–14
Delgado-V CA, French K (2015) Differential influence of urbanisation on coccidian infection in two passerine birds. Parasitol Res 114:2231–2235
Delgado-V CA, Pulgarín-R PC, Calderón-F D (2005) Análisis de egagrópilas del Búho rayado (*Asio clamator*) en la ciudad de Medellín. Ornitol Colomb 3:100–103
Dolbeer, RA (2015) Trends in reporting of wildlife strikes with civil aircraft and in identification of species struck under a primarily voluntary reporting system, 1990–2013. Tecnical report, U.S. Department of Agriculture Animal and Plant Health Inspection Service Wildlife Services. https://www.faa.gov/airports/airport_safety/wildlife/media/trends-in-wildlife-strike-reporting-1990-2013.pdf. Accessed 13 Apr 2017
Erickson WP, Johnson GD, Young DP Jr (2005) A summary and comparison of bird mortality from anthropogenic causes with an emphasis on collisions. Gen Tech Rep PSW-GTR-191. USDA Forest Service. Available via Treesearch. https://www.fs.fed.us/psw/publications/documents/psw_gtr191/psw_gtr191_1029-1042_erickson.pdf. Accessed 25 Apr 2017
Evans KL, Gaston KJ, Sharp SP et al (2009) Effects of urbanization on disease prevalence and age structure in blackbird *Turdus merula* populations. Oikos 118:774–782
Fokidis HB, Greiner EC, Deviche P (2008) Interspecific variation in avian blood parasites and haematology associated with urbanization in a desert habitat. J Avian Biol 39:300–310
Geue D, Partecke J (2008) Reduced parasite infestation in urban Eurasian blackbirds (*Turdus merula*): a factor favouring urbanization? Can J Zool 86:1419–1425
Gondim LSQ, Abe-Sandes K, Uzêda RS et al (2010) *Toxoplasma gondii* and *Neospora caninum* in sparrows (*Passer domesticus*) in the northeast of Brazil. Vet Parasitol 168:121–124
Habberfield MW, St. Clair CC (2016) Ultraviolet lights do not deter songbirds at feeders. J Ornithol 157:239–248
Hager SB, Cosentino BJ, McKay KJ (2012) Scavenging affects persistence of avian carcasses resulting from window collisions in an urban landscape. J Field Ornithol 83:203–211
Hamer SA, Lehrer E, Magle SB (2012) Wild birds as sentinels for multiple zoonotic pathogens along an urban to rural gradient in greater Chicago, Illinois. Zoonoses Public Health 59:355–364
Hernández-Lara C, González-García F, Santiago-Alarcon D (2017) Spatial and seasonal variation of avian malaria infections in five different land use types within a Neotropical montane forest matrix. Landsc Urban Plan 157:151–160

Johnson PTJ, Preston DL, Hoverman JT et al (2013) Biodiversity decreases disease through predictable changes in host community competence. Nature 494:230–233
Jones KE, Patel NG, Levy MA et al (2008) Global trends in emerging infectious diseases. Nature 451:990–993
Kilpatrick AM, Kramer LD, Jones MJ et al (2006) West Nile virus epidemics in North America are driven by shifts in mosquito feeding behavior. PLoS Biol 4:–e82
Klem D Jr, Farmer CJ, Delacretaz N et al (2009) Architectural and landscape risk factors associated with bird-glass collisions in an urban environment. Wilson J Ornithol 121:126–134
Klem D Jr, Saenger PG (2013) Evaluating the effectiveness of select visual signals to prevent bird-window collisions. Wilson J Ornithol 125:406–411
Komar N, Bessoff K, Diaz A et al (2012) Avian hosts of West Nile virus in Puerto Rico. Vector Borne Zoonotic Dis 12:47–54
Lafferty KD, Allesina S, Arim M et al (2008) Parasites in food webs: the ultimate missing links. Ecol Lett 11:533–546
Lima MR, Simpson L, Fecchio A et al (2010) Low prevalence of haemosporidian parasites in the introduced house sparrow (*Passer domesticus*) in Brazil. Acta Parasitol 55:297–303
Longcore T, Rich C, Mineau P et al (2012) An estimate of avian mortality at communication towers in the United States and Canada. PLoS One 7:e34025
Loss SR, Will T, Marra PP (2013) The impact of free-ranging domestic cats on wildlife of the United States. Nat Commun 4:1396
Loss SR, Will T, Loss SS et al (2014a) Bird-building collisions in the United States: estimates of annual mortality and species vulnerability. Condor 116:8–23
Loss SR, Will T, Marra PP (2014b) Estimation of bird-vehicle collision mortality on U.S roads. J Wildlife Manage 78:763–771
Loss SR, Will T, Marra PP (2015) Direct mortality of birds from anthropogenic causes. Annu Rev Ecol Evol Syst 46:99–120
Loyd KAT, Hernandez SM, Carroll JP et al (2013) Quantifying free-roaming domestic cat predation using animal-borne video cameras. Biol Conserv 160:183–189
MacGregor-Fors I, Schondube JE (2011) Gray vs. green urbanization: relative importance of urban features for urban bird communities. Basic Appl Ecol 12:372–381
MacGregor-Fors I, Avendaño-Reyes S, Bandala VM et al (2015) Multi-taxonomic diversity patterns in a neotropical green city: a rapid biological assessment. Urban Ecosyst 18:633–647
Machtans CS, Wedeles CHR, Bayne EM (2013) A first estimate for Canada of the number of birds killed by colliding with building windows. Avian Conserv Ecol 8:6
Manville AM II (2009) Towers, turbines, power lines, and buildings – Steps being taken by the U.S. Fish and Wildlife Service to avoid or minimize take of migratory birds at these structures. In: Rich TD, Arizmendi C, Demarest DW et al (eds) Tundra to tropics: connecting birds, habitats, and people. Proceedings of the 4th International Partners in Flight Conference. Partners in Flight, McAllen, p 262–272
Marcogliese DJ (2005) Parasites of the superorganism: are they indicators of ecosystem health? Int J Parasitol 35:705–716
Martin LB, Boruta M (2014) The impacts of urbanization on avian disease transmission and emergence. In: Gil D, Brumm H (eds) Avian urban ecology: behavioural and physiological adaptations. Oxford Univ. Press, Oxford, pp 116–128
Marzal A, Ricklefs RE, Valkiūnas G et al (2011) Diversity, loss, and gain of malaria parasites in a globally invasive bird. PLoS One 6:e21905
McCarthy RJ, Levine SH, Reed JM (2012) Estimation of effectiveness of three methods of feral cat population control by use of a simulation model. J Am Vet Med Assoc 243:502–511
Menacho-Odio RM (2015) Colisión de aves contra ventanas en Costa Rica: Conociendo el problema a partir de datos de museos, ciencia ciudadana y el aporte de biólogos. Zeledonia 19:10–21
Morales-Betoulle ME, Komar N, Panella NA et al (2013) West Nile virus ecology in a tropical ecosystem in Guatemala. Am J Trop Med Hyg 88:116–126

Nogales M, Vidal E, Medina FM et al (2013) Feral cats and biodiversity conservation: the urgent prioritization of island management. Bioscience 63:804–810

Ocampo-Peñuela N, Peñuela-Recio L, Ocampo-Durán Á (2016) Decals prevent bird-window collisions at residences: a successful case study from Colombia. Ornitol Colomb 15:94–101

Oliveira Cardoso C, do Nascimento Gomes D, Soares dos Santos AG et al (2014) Bird strike risk analysis at the international airport of Parnaíba, Piauí, Brazil. Ornitol Neotrop 25:179–193

Orduña-Villaseñor MV (2015) Dieta de gato doméstico *Felis silvestris catus* en el municipio de Morelia, Michoacán. MSc dissertation, Universidad Nacional Autónoma de México – Centro de Investigaciones en Ecosistemas (CIEco)

Ortega-Álvarez R, MacGregor-Fors I (2011) Dusting-off the file: a review of knowledge on urban ornithology in Latin America. Landsc Urban Plan 101:1–10

Oviedo S (2014) Estudio sobre preferencias de métodos utilizados para evitar el choque de aves contra puertas y ventanas de vidrio en Costa Rica. Programa Manejo de Recursos Naturales, Universidad Estatal a Distancia. Available via DOCZZ. http://doczz.es/doc/691615/colisi%C3%B3n-de-aves-contra-ventanas-en-costa-rica. Accessed 17 Apr 2017

Padilla LR, Santiago-Alarcon D, Merkel J et al (2004) Survey for *Haemoproteus* spp, *Trichomonas gallinae*, *Chlamydophila psittaci*, and *Salmonella* spp. in Galápagos Islands Columbiformes. J Zoo Wildl Med 35:60–64

Palinauskas V, Valkiūnas G, Bolshakov VC et al (2008) *Plasmodium relictum* (lineage P-SGS1): effects on experimentally infected passerine birds. Exp Parasitol 120:372–380

Poulin R (2007) Evolutionary ecology of parasites, 2nd edn. Princeton University Press, New Jersey

Poulin R, Morand S (2000) The diversity of parasites. Q Rev Biol 75:277–293

Roberts LS, Janovy J Jr (2009) Foundations of parasitology, 8th edn. McGraw-Hill, New York

Santiago-Alarcon D, Havelka P, Schaefer HM et al (2012) Bloodmeal analysis reveals avian *Plasmodium* infections and broad host preferences of *Culicoides* (Diptera: Ceratopogonidae) vectors. PLoS One 7:e31098

Santiago-Alarcon D, Carbó-Ramírez P, MacGregor-Fors I, Chávez-Zichinelli CA, Yeh PJ. Are cities that risky? An experimental test of lower avian malaria prevalence in an urban vs. a non-urban setting using a successful non-native urban invasive bird (in preparation)

Santiago-Alarcon D, Havelka P, Pineda E et al (2013) Urban forests as hubs for novel zoonosis: blood meal analysis, seasonal variation in *Culicoides* (Diptera: Ceratopogonidae) vectors, and avian haemosporidians. Parasitology 140:1799–1810

Santiago-Alarcon D, MacGregor-Fors I, Kühnert K et al (2016) Avian haemosporidian parasites in an urban forest and their relationship to bird size and abundance. Urban Ecosyst 19:331–346

Shochat E, Lerman SB, Anderies JM et al (2010) Invasion, competition, and biodiversity loss in urban ecosystems. Bioscience 60:199–208

Smallwood KS (2013) Comparing bird and bat fatality-rate estimates among north American wind-energy projects. Wildl Soc Bull 37:19–33

Stevens MHH (2009) A primer of ecology with R. Springer, New York

Suárez-Rodríguez M, López-Rull I, Macías Garcia C (2012) Incorporation of cigarette butts into nests reduces nest ectoparasite load in urban birds: new ingredients for an old recipe? Biol Lett 9:20120931

Suárez-Rodríguez M, Montero-Montoya RD, Macías Garcia C (2017) Anthropogenic nest materials may increase breeding costs for urban birds. Front Ecol Evol 5:4

van der Ree R, Heinze D, McCarthy M et al (2009) The ecology of roads in urban and urbanising landscapes. In: McDonell MJ, Hahs AK, Breuste JH (eds) Ecology of cities and towns: a comparative approach. Cambridge University Press, Cambridge, pp 185–196

van Heezik Y, Smyth A, Adams A et al (2010) Do domestic cats impose an unsustainable harvest on urban bird populations? Biol Conserv 143:121–130

Chapter 8
Urban Conservation: Toward Bird-Friendly Cities in Latin America

Augusto João Piratelli, Alexandre Gabriel Franchin, and Oscar H. Marín-Gómez

Abstract Urbanization is a major threat to biodiversity. Nevertheless, an important number of species has been recorded to live, and even thrive, within urban centers. Bird diversity has often been studied among greenspace networks, where most urban biodiversity is sheltered. Although a few studies have directly addressed urban bird conservation, they have shed important light on the directions and implications that evidence-based actions require for proper management and planning to occur in cities. Important international documents (e.g., *Rio+20,* FAO reports) have pinpointed the importance of developing greener cities in Latin America. Also, pioneer conservation ideas, such as *Archipelago Preserves*, have been conceived in Latin America as a way to plan conservation in advance where cities are predicted to grow. Birds are excellent bioindicators in urban areas for diverse reasons, as they quickly respond to habitat alterations and allow to monitor the ecological quality of urban areas. Nevertheless, for proper urban management and planning to occur in Latin America, a broader and robust evidence-based foundation is crucial, followed by interest and willingness from all the implied stakeholders for action to be made. Creating greener and biodiverse cities will not only benefit wildlife groups, including birds, but will certainly have a positive impact on people, making cities healthier and livable.

A.J. Piratelli (✉)
Departamento de Ciências Ambientais/CCTS, Universidade Federal de São Carlos, Campus Sorocaba, Rodovia João Leme doss Santos, Km 110, Itinga, 18052-780 Sorocaba, Sao Paulo, Brazil
e-mail: piratelli@ufscar.br

A.G. Franchin
Researcher collaborator - Laboratório de Ornitologia e Bioacústica, Instituto de Biologia, Universidade Federal de Uberlândia,
Rua Ceará s/n, Bloco 2D, Campus Umuarama, 38400-902 Uberlândia, Minas Gerais, Brazil

O.H. Marín-Gómez
Red de Ambiente y Sustentabilidad, Instituto de Ecología, A.C.
Carretera antigua a Coatepec 351, El Haya, Xalapa, 91070 Veracruz, Mexico

I. MacGregor-Fors, J.F. Escobar-Ibáñez (eds.), *Avian Ecology in Latin American Cityscapes*, DOI 10.1007/978-3-319-63475-3_8

Keywords Avian conservation • Citizen engagement • Habitat quality • Urban preserves • Urbanization intensity

8.1 Global Wildlife Sensitivity to Urbanization

Among the main consequences of urbanization, the implementation of nonsustainable processes in the exploitation of natural resources (Grimm et al. 2008; McDonnell and MacGregor-Fors 2016) and the endangerment of wildlife species (Czech et al. 2001; Maxwell et al. 2016) head the list. Many native species disappear from cities or shelter in large vegetation remnants because they cannot adjust to use and/or exploit urban resources and avoid the additional risks faced in cities (Sol et al. 2013; McDonnell and Hahs 2015; Chap. 6). Yet, some species are able to live, and even thrive, in urban areas, highlighting the importance of cities for conservation (Sanderson and Huron 2011; Aronson et al. 2014). Urban systems tend to favor species with greater phenotypic plasticity, such as generalists and species typical of edges and open areas that can adapt to anthropogenic selection pressures (González-Oreja 2011; see Chaps. 2, 3 and 6).

Since the 1970s, urban systems have been documented to support complex bird communities (Emlen 1974; Gavareski 1976; Lancaster and Rees 1979; Moore 1979; Clergeau et al. 1998 and references therein). Birds have been considered as an important biodiversity indicator group (also known as bioindicators) due to their high diversity, as well as the existence of robust methods to survey them and their stable taxonomy (Moreno et al. 2007). Together with the latter, birds have been documented to conform complex communities within urban centers, shifting along gradients of urbanization intensity and human activities (Blair 1996; Clergeau et al. 1998; Melles et al. 2003; Leveau and Leveau 2004; van Heezik et al. 2008; Ortega-Álvarez and MacGregor-Fors 2009; Villegas and Garitano-Zavala 2010; Biamonte et al. 2011; Reis et al. 2012; Clucas and Marzluff 2015; Silva et al. 2015).

Recently, Fischer et al. (2015) revisited Blair's (1996) categorization of wildlife responses to urbanization, concluding that there are species that avoid urban systems (but can dwell within lowly developed areas within cities; i.e., urban 'avoiders'), species that are common in urban areas but are mainly nonbreeders within them and depend on nonurban populations to maintain their dynamics (i.e., urban 'utilizers'), and species that are successful in exploiting urban resources and do not necessarily depend on nonurban populations to maintain their numbers in urban systems (i.e., urban 'dwellers'). This is in agreement with the idea that urban systems pose an ecological barrier for biodiversity, with birds not being an exception (Croci et al. 2008; MacGregor-Fors 2010; Puga-Caballero et al. 2014). Nevertheless, the barrier is 'semipermeable', allowing the entrance of those species that can reach the urban system, use their resources, and survive to their hazards (Emlen 1974).

Although local factors have been suggested to override regional ones (Evans et al. 2009), scale has shown to be determining for urban birds. Also, studies have identified several predictors related to the ability of species to live and/or thrive in urban systems: (i) taxonomical identity (reviewed in Chap. 3); (ii) feeding behavior and requirements (Faeth et al. 2005; Chaps. 3 and 6); (iii) biogeographical origin (González-Oreja 2011); (iv) spatial distribution (MacGregor-Fors and Ortega-Álvarez 2011); (v) physiology (Chávez-Zichinelli et al. 2013); (vi) ecological plasticity (Kenney and Knight 1992); (vii) adaptation (Bonier et al. 2007; Chap. 6); and (viii) sociability, sedentary lifestyle, reproductive cycles, and nesting habits (Kark et al. 2007).

8.2 Avian Conservation in Urban Systems

8.2.1 General Perspectives

Conservation biology in urban areas has been questioned in the past due to the relative impoverishment of biodiversity caused by urbanization. Yet, there are several reasons behind the recent interest in pursuing biological conservation in urban centers, some of which were brilliantly highlighted by Sanderson and Huron (2011) in an editorial of *Conservation Biology*. As they note, the most intuitive reason behind people pushing toward conservation practices and plans in urban centers is because most of us are urbanites. Also, conservation biology in urban areas challenges its concept and requires the development of new frameworks that leap forward toward the accomplishment of biological conservation in human-modified systems and landscapes (Sanderson and Huron 2011).

Greener urban scenarios tend to harbor more bird species (Chace and Walsh 2006; González-García et al. 2014). In fact, the size and connectivity of greenspaces in cities are recurrent topics in the urban bird conservation literature (e.g., Shanahan et al. 2011). In essence, greenspaces may serve as 'sources' for certain species, which can disperse to the surrounding human settlements (van Heezik et al. 2008). This leads to the controversial weighting of the land-sharing/land-sparing paradigms, a debate formulated for agriculture production, but with parallels in urbanization models (Lin and Fuller 2013). In the first case, the whole urban territory would be developed heterogeneously by mixing buildings and greenspaces. For the land-sparing model, large blocks of either intensely urbanized or greenspaces predominate in a broader scale. There are different responses of biodiversity and human well-being for each model, and understanding them may result in functional, well-planned and sustainable cities, thus reducing the overall impact of urban sprawl on biodiversity (Soga et al. 2014).

Several questions have been addressed and have set the foundations to build a framework to generate evidence-based solutions toward healthier cities. Greenspace size is certainly a key element, as larger areas tend to increase spatial and environmental heterogeneity, and thus shelter greater bird diversity (Fernández-Juricic 2000a; Platt and Lill 2006; Palmer et al. 2008; Evans et al. 2009; Schütz and Schulze 2015; Chang and Lee 2016; MacGregor-Fors et al. 2016). Another factor that has been positively related with urban bird diversity is the age of native vegetation patches and remnants (Fernández-Juricic 2000a). Greenspaces are priority areas for conservation (Palmer et al. 2008), providing the basis toward long-term conservation (Schütz and Schulze 2015; Chang and Lee 2016). In this sense, large greenspaces must be prioritized as they have been shown to offer a trade-off against the deleterious effects of built cover and the discontinuous distribution of potential habitats for native species (Fernández-Juricic 2000a, b; Evans et al. 2009; MacGregor-Fors et al. 2016).

John M. Marzluff (2014) named the 'suburbs' of large cities 'subirdia' due to the surprising biological avian diversity they shelter. In his book, Marzluff (2014) highlights that some urban environments can house and maintain biodiversity if we provide three key elements: (i) suitable habitat, including sites for shelter and breeding, food, and water sources; (ii) safety, reducing bird mortality caused by both direct and indirect human causes (see Chap. 7 for a thorough review on this topic); and (iii) engagement, involvement of residents, including environmental education focused on teens and children. Another iconic book on urban conservation, by Mark Hostetler (2012), underlines the crucial role that engagement of residents as a cornerstone strategy for conservation; otherwise, the urban population would not be aware of, understand, or participate in any actions. Although authorities generally implement conservation policies, they do not necessarily succeed in cities without public participation, highlighting the importance of involving all urban stakeholders in decision-making and practice. For instance, if each household owner in a neighborhood decided to design and manage their lot to integrate and house native plants and animals, the entire area would be benefited by individual decisions (Hostetler 2012).

8.2.2 *Perspectives from Urban Latin America*

As noted above, there are several reasons behind the development of urban conservation strategies. Yet, there are some biodiverse regions where this must be prioritized, such as Latin America (Hedblom and Murgui 2017). With its territory sheltering important global biodiversity, four out of the five richest countries in terms of bird diversity are Latin American (i.e., Colombia, Peru, Brazil, Ecuador; BirdLife International 2013). Also, Latin America is a region where poverty, inequality, and urbanization rates meet with ~30% of the global hot spots for conservation priorities (Myers et al. 2000; Pauchard and Barbosa 2013).

Although urban bird communities share several hyperabundant species across the globe (Sol et al. 2013; Aronson et al. 2014), leading to the idea that biotic homogenization occurs in urban areas (McKinney 2006; Devictor et al. 2007), global trends are not clear (Sol et al. 2013; Aronson et al. 2014; see Chap. 9 for a

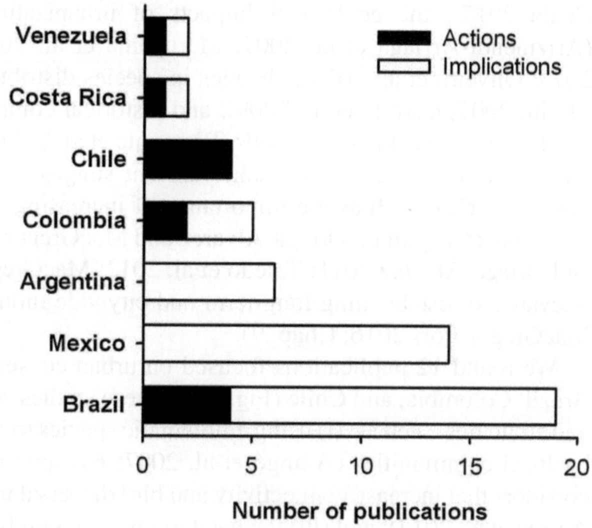

Fig. 8.1 Number of publications focused on conservation implications or actions in urban Latin America

thorough discussion on this topic). Indeed, avian diversity in urban Latin America has shown to be high when compared with that recorded in other well-studied regions. For example, González-García et al. (2014, 2016) report 341 species for the city of Xalapa (Mexico) in the last three decades, while Biamonte et al. (2011) report more than 200 species for San José (Costa Rica), and Franchin (2009) reports more than 150 species in seven Brazilian urban centers (i.e., Londrina, Marabá, Pirassununga, Porto Alegre, Rio Claro, São Paulo, Uberlândia).

Despite the importance of the topic, bird conservation studies from the urban Latin American scientific literature are scarce. In fact, when searching for publications that have dealt with avian conservation in urban Latin America in the Web of Science (www.webofknowledge.com) and Scopus (www.scopus.com) platforms using the keyword combination: 'bird' AND 'urban' AND 'conservation' AND 'each Latin American country' (see Chaps. 2 and 3 for a list of countries), we only retrieved 76 documents from eight countries, with Brazil and Mexico heading the list (Fig. 8.1). Afterward, we filtered the studies that focused on conservation implications or actions. We considered conservation implication studies as those focused on the ecological effects of urbanization on birds that report conservation suggestions based on their findings, while classified those presenting conservation management plans, environmental education activities, and/or community-based conservation efforts as conservation action studies (see Chaps. 3, 4 and 5 for a detailed review and analysis of the conservation implications for bird-friendly cities in urban Latin America).

Studies with urban conservation implications (40 publications) have focused on highly relevant topics, including the effects of environmental characteristics and anthropogenic changes on bird community structure and species diversity (Ortega-Álvarez and MacGregor-Fors 2009; Ruelas Inzunza and Aguilar Rodríguez 2010; Cruz and Piratelli 2011; Domínguez-López and Ortega-Álvarez 2014; Sanz and

Caula 2015), the ecological impacts of urbanization in mutualistic networks (Arizmendi-Arriaga et al. 2007; Maruyama et al. 2012; Ferreira and Consolaro 2013; Oliveira et al. 2013), changes in species distribution patterns (Carvalho and Marini 2007; Carrete et al. 2009), and historical community approaches reporting local extinctions of resident birds (Biamonte et al. 2011; Moura et al. 2014). Among the reviewed studies, some make relevant suggestions for urban conservation in Latin America, such as the importance of increasing environmental heterogeneity and connectivity (e.g., Ortega-Álvarez and MacGregor-Fors 2009; MacGregor-Fors and Ortega-Álvarez 2011; Toledo et al. 2012; MacGregor-Fors et al. 2016) and the relevance of establishing long-term and citywide monitoring (Escobar-Ibáñez and MacGregor-Fors 2016; Chap. 9).

We found 12 publications focused on urban conservation actions, mainly from Brazil, Colombia, and Chile (Fig. 8.1). These studies address a wide array of topics and strategies, such as: (i) using emblematic species to promote conservation actions by local communities (Arango et al. 2007; Azevedo et al. 2012); (ii) considering corridors that increase connectivity and bird dispersal in urban ecosystems (Trujillo-Acosta et al. 2017); and (iii) the need to limit the number of beach tourists to reduce their impact on migratory shorebirds (Cestari 2015). These studies highlight the complexity of applying conservation actions in urban areas, identifying several limitations for them to be successful (e.g., lack of strategies to promote urbanite environmental education, the separation of humans from nature, the unwillingness of all stakeholders to embrace biodiverse urban scenarios; Turner et al. 2004; Bezerra et al. 2012; Cestari 2015).

8.3 Conclusions and Future Directions

Our knowledge regarding the adaptiveness of birds to urban areas is still consolidating, although there is important evidence of avian plasticity in these systems (McDonnell and Hahs 2015; Chap. 6). Several traits related to the capacity to adapt and/or respond (positively or negatively) to urbanization have been identified, including their evolutionary history (González-Oreja 2011), morphology (Evans et al. 2009), physiology (Chavez-Zichinelli et al. 2013), feeding ecology (Escobar-Ibáñez and MacGregor-Fors 2015) (Fig. 8.2), reproductive habits (Chace and Walsh 2006) (Fig. 8.3), sociability (Kark et al. 2007), and behavior (Seress and Liker 2015). Thus, their capacity to quickly respond to human disturbances, from individuals to communities, together with them being highly charismatic and generally well accepted by urbanites, makes them an ideal informative wildlife group for urban conservation.

Given the differential response of urban 'dwelling' birds to urbanization (as described by Fischer et al. 2015) and that rich native bird communities are closely related to large urban greenspaces (see Chaps. 3 and 5), which can include residential and/or commercial components (MacGregor-Fors et al. 2009), conservation strategies based on organisms like birds in urban areas should focus on the maximi-

Fig. 8.2 Chestnut-eared Aracari (*Pteroglossus castanotis*) feeding on the fruit of a palm tree in the city of Uberlândia, Brazil (Photo: AGF)

zation of suitable conditions and resources at different spatial scales (Chamberlain et al. 2009; Gagné et al. 2016). To migrate toward livable, biodiverse, and sustainable cities, urban management and planning is imperative (McDonnell and MacGregor-Fors 2016). In this sense, urban management needs to focus on both the urban matrix and greenspaces (see Chap. 5 for a review and discussion regarding both 'gray' and 'green' urban infrastructure). Due to the biodiversity that greenspaces shelter, there is a generalized bias of urban ecology knowledge and focus regarding conservation issues toward them. Yet, all urban scenarios need to be included in both management and planning strategies to generate favorable environments for both wildlife and urbanites (Hostetler and Knowles-Yanez 2003; Escobar-Ibáñez and MacGregor-Fors 2015). Besides private gardens, which have shown to be crucial for wildlife groups along urban greenspace networks (Cannon et al. 2005; Daniels and Kirkpatrick 2006; Andersson and Colding 2014; Belaire et al. 2016), all other urban spaces, such as sidewalks and median strips along streetscapes (Fernández-Juricic 2000b; Chamberlain et al. 2009; Carbó-Ramírez and Zuria 2011), ought to be considered when managing and planning cities(Matarazzo-Neuberger 1995; Donnelly and Marzluff 2006; Braga et al. 2010; Falfán and

Fig. 8.3 Great Kiskadee (*Pitangus sulphuratus*) nest with nestlings showing the use of artificial components in the campus of the Federal University of Uberlândia, Brazil (Photo: AGF)

MacGregor-Fors 2016). New creative solutions, some of which are often accidental, need to be explored hand in hand with potential wildlife adaptations, such as native Peach-fronted Parakeets (*Eupsittula aurea*) nesting in air-conditioning units in Campo Grande, Brazil (Souza 2016); Azure-crowned Hummingbirds (*Amazilia cyanocephala*) successfully nesting on open-sky metallic structures of the telephone wiring in Xalapa, Mexico (Escobar-Ibáñez and MacGregor-Fors 2015), and Rufous Horneros (*Furnarius rufus*) nesting on high-voltage electricity poles in the state of Santa Catarina, Brazil (Efe and Filippini 2006).

Regarding urban greenspaces, many management and planning strategies have been identified as crucial for creating bird-friendly cities, many supported in Latin American studies, among which the following head the list: (i) diversifying the use of urban greenspaces, including conservation goals (Lagoa 2008; Carbó-Ramírez and Zuria 2011; Domínguez-López and Ortega-Álvarez 2014; Carbone et al. 2015; Sanz and Caula 2015; Silva et al. 2015); (ii) redesigning urban greenspaces considering complex citywide networks (Dredge 1995; Efe et al. 2001; Bargos and Matias 2011; Wendel et al. 2012; Andrade et al. 2013; MacGregor-Fors et al. 2016; Trujillo-Acosta et al. 2017); (iii) promoting native vegetation structure and composition in existing greenspaces, limiting the use of exotic species (although this is a complex issue; see Chaps. 5 and 9) (MacGregor-Fors 2008; Palmer et al. 2008; Ziller and Dechoum 2013; Falfán and MacGregor-Fors 2016); (iv) reconsidering vegetation management strategies and regimes (Sandström et al. 2006; Chang and Lee 2016; Uras et al. 2014); and (v) using low-maintenance trees that provide food and nesting resources (Mendonça and Anjos 2005; Castro et al. 2010; Oliveira et al. 2013).

Additionally, and closely related to urban planning toward wildlife-friendly cities, it is fundamental to protect target native vegetation patch networks where cities are predicted to sprawl. Thus, the use of concepts like Gonzalo Halffter's *Archipelago Preserves* (2007) and other novel biodiversity protection strategies need to be considered and implemented by urban planners and developers together with local and regional authorities.

As reviewed in Chap. 5, all types of urban infrastructure need to be assessed if we aim to achieve livable, sustainable, and bird-friendly cities. We would like to emphasize on the importance of water bodies and their related vegetation, which are often central to the successful management of urban bird diversity. For example, a four-year restoration of a riparian forest dominated by exotic grasses in Puerto Rico has resulted in microclimate stabilization, reduction of herbaceous cover, increased litter fall, and colonization by plant and animal species (Ruiz-Jaén and Aide 2006). In Brazil, there are laws that protect riverbanks as permanent protected areas (e.g., Law 771, Article 1st, 2nd, II, September 15, 1965; Brazil), established to preserve water resources, biodiversity, and human welfare (PRB 1965). Undoubtedly, the protection and restoration of rivers, lakes, and other urban water bodies would also likely favor biodiversity in urban areas where present or manageable (Clergeau et al. 2001; Cruz and Piratelli 2011; Rosselli and Stiles 2012a, b), serving as stepping stones for migratory species, for instance (Fig. 8.4).

For urban management and planning strategies to promote bird-friendly cities in Latin America, we need a robust evidence-based knowledge foundation followed by interest and willingness from all the implied stakeholders for action to be made (MacGregor-Fors 2015). Evidently, the implementation of protected areas and proper management of urban greenspaces require economic resources (Burkhalter et al. 2016); yet, active interest of the implied decision-makers often overrides the importance of investments. Important international documents, such as the *Rio+20–The Future we Want* (Jordán 2013) and a recent report from the Food and Agriculture Organization of the United Nations (FAO 2014) have pinpointed the importance of urban management and planning, seeking sustainable development and greener cities in Latin America. Furthermore, other initiatives, such as the *Latin American Green City Index* (Rodríguez Tejerina 2015), have been created to evaluate the environmental performance of major Latin American cities. The aim, in this case, was to identify the most sustainable cities, according to their practices, considering energy, solid waste management, and water and air quality, among other factors (Siemens 2010). Similarly, *The Urban Management Program for Latin America and Caribbean* (UMP-LAC 2001) aims to achieve environmental sustainability by encouraging urban agriculture and controlling river pollution and solid waste management.

Furthermore, it is crucial to include citizens and promote environmental education to develop successful conservation actions for urban biodiversity. A remarkable example of a citizen-based network of information is the eBird platform, which encourages amateur and professional birdwatchers, as well as ornithologists and ecologists, to report bird observations. The information gathered in the platform is without doubt increasing our knowledge on bird distribution and natural history to

Fig. 8.4 Urban pond in Sorocaba (Brazil) hosting a migratory population of the Roseate Spoonbill (*Platalea ajaja*) (Photo: AJP)

the extent that eBird is considered one of the largest and growing biodiversity data sources worldwide (Sullivan et al. 2014). This kind of effort does not only provide information, but also has an important role in bridging the gap between urbanites, urban ecology, and decision-making (Sullivan et al. 2016). Moreover, local birding and ornithological groups along Latin America work together in several environmental education networks to promote the value of conservation of urban systems.

Yet, many Latin American cities are still growing at unprecedented rates without proper environmental management or planning (Cerrutti and Bertoncello 2003; Cohen 2004; Lankao 2007; Schneider and Woodcock 2008). As stated earlier, birds are an excellent group for studying ecological patterns in cities, even more in those where they are highly diverse, and can reflect the ecological quality of urban systems in urban Latin America (Reynaud and Thioulouse 2000; Alberti 2005; MacGregor-Fors et al. 2009). Creating greener and biodiverse cities will not only benefit birds and wildlife in general, but will have a positive impact on citizens, including health and ecoservices (Gómez-Baggethun et al. 2013; Elmqvist et al. 2016).

Acknowledgments The authors thank the reviewer and editors for their valuable comments and suggestions. OHMG acknowledges the doctoral scholarship and financial support provided by the National Council of Science and Technology (CONACYT 417094) and the PhD Program of the Instituto de Ecología, A.C. (INECOL).

References

Alberti M (2005) The effects of urban patterns on ecosystem function. Int Reg Sci Rev 28:168–192

Andersson E, Colding J (2014) Understanding how built urban form influences biodiversity. Urban For Urban Gree 13:221–226

Andrade GI, Remolina F, Wiesner D (2013) Assembling the pieces: a framework for the integration of multi-functional ecological main structure in the emerging urban region of Bogotá, Colombia. Urban Ecosyst 16:723–739

Arango X, Rozzi R, Massardo F et al (2007) Descubrimiento e implementación del pájaro carpintero gigante (*Campephilus magellanicus*) como especie carismática: una aproximación biocultural para la conservación en la Reserva de Biosfera Cabo de Hornos. Magallania 35:71–88

Arizmendi-Arriaga MC, Constanza MS, Lourdes J et al (2007) Effect of the presence of nectar feeders on the breeding success of *Salvia mexicana* and *Salvia fulgens* in a suburban park near Mexico City. Biol Conserv 136:155–158

Aronson MFJ, La Sorte FA, Nilon CH et al (2014) A global analysis of the impacts of urbanization on bird and plant diversity reveals key anthropogenic drivers. Proc R Soc Lond B Biol Sci 281:20133330

Azevedo CS, Silva KS, Ferraz JB et al (2012) Does people's knowledge about an endangered bird species differ between rural and urban communities? The case of the Greater Rhea (*Rhea americana*, Rheidae) in Minas Gerais, Brazil. Rev Bras Ornitol 20:8–18

Bargos DC, Matias LF (2011) Áreas verdes urbanas: um estudo de revisão e proposta conceitual. REVSBAU 6:172–188

Belaire JA, Westphal LM, Minor ES (2016) Different social drivers, including perceptions of urban wildlife, explain the ecological resources in residential landscapes. Landsc Ecol 31:401–413

Bezerra DMM, de Araujo HFP, Alves RRN (2012) Captura de aves silvestres no semiárido brasileiro: técnicas cinegéticas e implicações para conservação. Trop Conserv Sci 5:50–66

Biamonte E, Sandoval L, Chacón E et al (2011) Effect of urbanization on the avifauna in a tropical metropolitan area. Landsc Ecol 26:183–194

BirdLife International (2013) Birds are found almost everywhere in the world, from the poles to the equator. In: Data zone. BirdLife International. http://datazone.birdlife.org/sowb/casestudy/birds-are-found-almost-everywhere-in-the-world-from-the-poles-to-the-equator. Accessed 3 May 2017

Blair RB (1996) Land use and avian species diversity along an urban gradient. Ecol Appl 6:506–519

Bonier F, Martin PR, Wingfield JC (2007) Urban birds have broader environmental tolerance. Biol Lett 3:670–673

Braga TV, Zanzini ACDS, Cerboncini RAS et al (2010) Avifauna em praças da cidade de Lavras (MG): riqueza, similaridade e influência de variáveis do ambiente urbano. Rev Bras Ornitol 18:26–33

Burkhalter JC, Lockwood JL, Maslo B et al (2016) Effects of cost metric on cost-effectiveness of protected-area network design in urban landscapes. Conserv Biol 30:403–412

Cannon AR, Chamberlain DE, Toms MP et al (2005) Trends in the use of private gardens by wild birds in Great Britain 1995-2002. J Appl Ecol 42:659–671

Carbone AS, Coutinho SMV, Tomerius S et al (2015) The management of green areas in the municipality of São Paulo: advances and limitations. Ambient Soc 18:195–214

Carbó-Ramírez P, Zuria I (2011) The value of small urban greenspaces for birds in a Mexican city. Landsc Urban Plan 100:213–222

Carrete M, Tella JL, Blanco G et al (2009) Effects of habitat degradation on the abundance, richness and diversity of raptors across Neotropical biomes. Biol Conserv 142:2002–2011

Carvalho CEA, Marini MA (2007) Distribution patterns of diurnal raptors in open and forested habitats in south-eastern Brazil and the effects of urbanization. Bird Conserv Int 17:367–380

Castro WS, Franchin AG, Marçal Junior O (2010) Reprodução de *Glaucidium brasilianum* (Gmelin, 1788) em área urbana deUberlândia, Minas Gerais. Rev Bras Ornitol 18:55–58

Cerrutti M, Bertoncello R (2003) Urbanization and internal migration patterns in Latin America. Paper presented at conference on African migration in comparative perspective, Johannesburg, 4–7 June, 2003

Cestari C (2015) Coexistence between Nearctic-Neotropical migratory shorebirds and humans on urban beaches of the Southern Hemisphere: a current conservation challenge in developing countries. Urban Ecosyst 18:285–291

Chace JF, Walsh JJ (2006) Urban effects on native avifauna: a review. Landsc Urban Plan 74:46–69

Chamberlain DE, Cannon AR, Toms MP et al (2009) Avian productivity in urban landscapes: a review and meta-analysis. Ibis 151:1–18

Chang HY, Lee YF (2016) Effects of area size, heterogeneity, isolation, and disturbances on urban park avifauna in a highly populated tropical city. Urban Ecosyst 19:257–274

Chávez-Zichinelli CA, MacGregor-Fors I, Quesada J et al (2013) How stressed are birds in an urbanizing landscape? Relationships between the physiology of birds and three levels of habitat alteration. Condor 115:84–92

Clergeau P, Savard JPL, Mennechez G et al (1998) Bird abundance and diversity along an urban-rural gradient: a comparative study between two cities on different continents. Condor 100:413–425

Clergeau P, Jokimäki J, Savard JPL (2001) Are urban bird communities influenced by the bird diversity of adjacent landscapes? J Appl Ecol 38:1122–1134

Clucas B, Marzluff JM (2015) A cross-continental look at the patterns of avian species diversity and composition across an urbanisation gradient. Wildl Res 42:554–562

Cohen B (2004) Urban growth in developing countries: a review of current trends and a caution regarding existing forecasts. World Dev 32:23–51

Croci S, Butet A, Clergeau P (2008) Does urbanization filter birds on the basis of their biological traits? Condor 110:223–240

Cruz BB, Piratelli AJ (2011) Avifauna associated to an urban extend of the Sorocaba River, Southeastern Brazil. Biota Neotrop 11:255–264

Czech B, Devers PK, Krausman PR (2001) The relationship of gender to species conservation attitudes. Wildl Soc B 29:187–194

Daniels GD, Kirkpatrick JB (2006) Does variation in garden characteristics influence the conservation of birds in suburbia? Biol Conserv 133:326–335

Devictor V, Julliard R, Couvet D et al (2007) Functional homogenization effect of urbanization on bird communities. Conserv Biol 21:741–751

Domínguez-López ME, Ortega-Álvarez R (2014) The importance of riparian habitats for avian communities in a highly human-modified Neotropical landscape. Rev Mex Biodivers 85:1217–1227

Donnelly R, Marzluff JM (2006) Relative importance of habitat quantity, structure, and spatial pattern to birds in urbanizing environments. Urban Ecosyst 9:99–117

Dredge D (1995) Sustainable rapid urban expansion: the case of Xalapa, Mexico. Habitat Int 19:317–329

Efe MA, Filippini A (2006) Nidificação do joão-de-barro, *Furnarius rufus* (Passeriformes, Furnariidae) em estruturas de distribuição de energia elétrica em Santa Catarina. Ornithologia 1:121–124

Efe MA, Mohr LV, Bugoni L et al (2001) Inventário e distribuição da avifauna do Parque Saint'Hilaire, Viamão, Rio Grande do Sul, Brasil. Tangara 1:12–25

Elmqvist T, Zipperer W, Güneralp B et al (2016) Urbanization, habitat loss, biodiversity decline: solution pathways to break the cycle. In: Seto KC, Solecki WD, Griffith CA (eds) The Routledge handbook of urbanization and global environmental change. Routledge, London, pp 139–151

Emlen JT (1974) An urban bird community in Tucson, Arizona: derivation, structure, regulation. Condor 76:184–197

Escobar-Ibáñez JF, MacGregor-Fors I (2015) On a tightrope: use of open sky urban telephone wires by Azure-crowned Hummingbirds (*Amazilia cyanocephala*) for nesting. Wilson J Ornithol 127:297–302

Escobar-Ibáñez JF, MacGregor-Fors I (2016) Peeking into the past to plan the future: assessing bird species richness in a neotropical city. Urban Ecosyst 19:657–667

Evans KL, Newson SE, Gaston KJ (2009) Habitat influences on urban avian assemblages. Ibis 151:19–39

Faeth SH, Warren PS, Shochat E et al (2005) Trophic dynamics in urban communities. BioScience 55:399–409

Falfán I, MacGregor-Fors I (2016) Woody neotropical streetscapes: a case study of tree and shrub species richness and composition in Xalapa. Madera Bosques 22:95–110

Fernández-Juricic E (2000a) Bird community composition patterns in urban parks of Madrid: the role of age, size and isolation. Ecol Res 15:373–383

Fernández-Juricic E (2000b) Avifaunal urban use of wooded streets in an landscape. Conserv Biol 14:513–521

Ferreira MC, Consolaro H (2013) Fenologia e síndromes de polinização e dispersão de espécies de sub- bosque em um remanescente florestal urbano no Brasil Central. Bioscience 29:1708–1720

Fischer JD, Schneider SC, Ahlers AA et al (2015) Categorizing wildlife responses to urbanization and conservation implications of terminology. Conserv Biol 29:1246–1248

Food and Agriculture Organization (FAO) (2014) Growing greener cities in Latin America and the Caribbean: an FAO report on urban and peri-urban agriculture in the region. Food and Agriculture Organization of the United Nations. http://www.fao.org/3/a-i3696e.pdf. Accessed 3 May 2017

Franchin AG (2009) Avifauna em áreas urbanas brasileiras, com ênfase em cidades do Triângulo Mineiro/Alto Paranaíba Triângulo Mineiro/Alto Paranaíba. Universidade Federal de Uberlândia, PhD Dissertation

Gagné A, Sherman PJ, Singh KK et al (2016) The effect of human population size on the breeding bird diversity of urban regions. Biodivers Conserv 25:653–671

Gavareski CA (1976) Relation of park size and vegetation to urban bird populations in Seattle, Washington. Condor 78:375–382

Gómez-Baggethun E, Gren A, Barton D et al (2013) Urban ecosystem services. In: Elmqvist T, Fragkias M, Goodness J et al (eds) Urbanization, biodiversity and ecosystem services: challenges and opportunities. Springer, London, pp 175–251

González-García F, Straub R, Lobato García JA et al (2014) Birds of a Neotropical green city: an up-to-date review of the avifauna of the city of Xalapa with additional unpublished records. Urban Ecosyst 17:991–1012

González-García F, Straub R, Lobato García JA et al (2016) Nuevos registros y notas adicionales comentadas sobre la avifauna de la ciudad de Xalapa, Veracruz, México. Acta Zool Mex (ns) 32:253–269

González-Oreja JA (2011) Birds of different biogeographic origins respond in contrasting ways to urbanization. Biol Conserv 144:234–242

Grimm NB, Faeth SH, Golubiewski NE et al (2008) Global change and the ecology of cities. Science 319:756–760

Halffter G (2007) Reservas archipiélago: un nuevo tipo de área protegida. In: Halffter G, Guevara S, Melic A (eds) Hacia una cultura de conservación de la diversidad biológica. m3m: Monografías Tercer Milenio, Zaragoza, pp 281–286

Hedblom M, Murgui E (2017) Urban bird research in a global perspective. In: Mugui E, Hedblom M (eds) Ecology and conservation of birds in urban environments. Springer, Cham, pp 3–10

van Heezik Y, Smyth A, Mathieu R (2008) Diversity of native and exotic birds across an urban gradient in a New Zealand city. Landsc Urban Plan 87:223–232

Hostetler M (2012) The green leap: a primer for conserving biodiversity in subdivision development. University of California Press, Berkeley

Hostetler M, Knowles-Yanez K (2003) Land use, scale, and bird distributions in the Phoenix metropolitan area. Landsc Urban Plan 62:55–68
Jordán R (2013) Urban sustainability in Latin America and the Caribbean. CSTD 2012-2013 Inter-Sessional Panel, January 2013, Lima, Peru. http://unctad.org/meetings/en/Presentation/cstd2013_Jordan.pdf. Accessed 3 May 2017
Kark S, Iwaniuk A, Schalimtzek A et al (2007) Living in the city: can anyone become an 'urban exploiter'? J Biogeogr 34:638–651
Kenney SP, Knight RL (1992) Flight distances of Black-billed Magpies in different regimes of human density and persecution. Condor 94:545–547
Lagoa MHB (2008) O parque da água branca: o manejo sustentável de uma floresta urbana. Universidade de São Paulo, MSc Thesis
Lancaster RK, Rees WE (1979) Bird communities and the structure of urban habitats. Can J Zool 57:2358–2368
Lankao PR (2007) Are we missing the point? Particularities of urbanization, sustainability and carbon emissions in Latin American cities. Environ Urban 19:159–175
Leveau LM, Leveau CM (2004) Comunidades de aves en un gradiente urbano de la ciudad de Mar del Plata. El Hornero (B. Aires) 19:13–21
Lin BB, Fuller RA (2013) Sharing or sparing ? How should we grow the world's cities? J Appl Ecol 50:1161–1168
MacGregor-Fors I (2008) Relation between habitat attributes and bird richness in a western Mexico suburb. Landsc Urban Plan 84:92–98
MacGregor-Fors I (2010) How to measure the urban-wildland ecotone: redefining 'peri-urban' areas. Ecol Res 25:883–887
MacGregor-Fors I (2015) 10 scientists and 10 practitioners walk into a bar...what would they talk about? How can research and knowledge generation be co-created to better support practitioners and evidence-based decision making? In: Maddox D (ed) Global roundtable. The nature of cities. https://www.thenatureofcities.com/2015/06/30/10-scientists-and-10-practitioners-walk-into-a-barwhat-would-they-talk-about-how-can-research-and-knowledge-generation-be-co-created-to-better-support-practitioners-and-evidence-based-decisi/. Accessed 3 May 2017
MacGregor-Fors I, Ortega-Álvarez R (2011) Fading from the forest: bird community shifts related to urban park site-specific and landscape traits. Urban For Urban Green 10:239–246
MacGregor-Fors I, Ortega-Álvarez R, Schondube JE (2009) On the ecological quality of urban systems: an ornithological perspective. In: Graber DS, Birmingham KA (eds) Urban planning in the 21st century. Nova Science Publishing, New York, pp 51–66
MacGregor-Fors I, Escobar F, Rueda-Hernández R et al (2016) City "green" contributions: the role of urban greenspaces as reservoirs for biodiversity. Forests 7:1–14
Maruyama PK, Mendes-Rodrigues C, Alves-Silva E et al (2012) Parasites in the neighbourhood: interactions of the mistletoe *Phoradendron affine* (Viscaceae) with its dispersers and hosts in urban areas of Brazil. Flora 207:768–773
Marzluff JM (2014) Welcome to Subirdia: sharing our neighborhoods with wrens, robins, woodpeckers, and other wildlife. Yale University Press, New Haven
Matarazzo-Neuberger WM (1995) Comunidade de aves de cinco parques e praças da Grande São Paulo, Estado de São Paulo. Ararajuba 3:13–19
Maxwell SL, Fuller RA, Brooks TM et al (2016) The ravages of guns, nets and bulldozers. Nature 536:143–145
McDonnell MJ, Hahs AK (2015) Adaptation and Adaptedness of organisms to urban environments. Annu Rev Ecol Evol Syst 46:261–280
McDonnell MJ, MacGregor-Fors I (2016) The ecological future of cities. Science 352:936–938
McKinney ML (2006) Urbanization as a major cause of biotic homogenization. Biol Conserv 127:247–260
Melles S, Glenn S, Martin K (2003) Urban bird diversity and landscape complexity: species-environment associations along a multiscale habitat gradient. Conserv Ecol 7:5
Mendonça LB, Anjos L (2005) Beija-flores (Aves, Trochilidae) e seus recursos florais em uma área urbana do Sul do Brasil. Rev Bras Zool 22:51–59

Moore PD (1979) Urban ecology. Nature 282:13–14
Moreno CE, Sánchez-Rojas G, Pineda E et al (2007) Shortcuts for biodiversity evaluation: a review of terminology and recommendations for the use of target groups, bioindicators and surrogates. Int J Environ Health Res 1:71–86
Moura NG, Lees AC, Aleixo A et al (2014) Two hundred years of local avian extinctions in Eastern Amazonia. Conserv Biol 28:1271–1281
Myers N, Mittermeier RA, Mittermeier CG et al (2000) Biodiversity hotspots for conservation priorities. Nature 403:853–845
Oliveira DSF, Franchin AG, Marcal Junior O (2013) Availability of *Michelia champaca* L. (Magnoliaceae) fruits and its consumption by birds in the urban area of Uberlandia, state of Minas Gerais, Brazil. Biosci J 29:2053–2065
Ortega-Álvarez R, MacGregor-Fors I (2009) Living in the big city: effects of urban land-use on bird community structure, diversity, and composition. Landsc Urban Plan 90:189–195
Palmer GC, Fitzsimons JA, Antos MJ et al (2008) Determinants of native avian richness in suburban remnant vegetation: implications for conservation planning. Biol Conserv 141:2329–2341
Pauchard A, Barbosa O (2013) Regional assessment of Latin America: rapid urban development and social economic inequity threaten biodiversity hotspots. In: Elmqvist T, Fragkias M, Goodness J et al (eds) Urbanization, biodiversity and ecosystem services: challenges and opportunities. Springer, London, pp 589–608
Platt A, Lill A (2006) Composition and conservation value of bird assemblages of urban 'habitat islands': do pedestrian traffic and landscape variables exert an influence? Urban Ecosyst 9:83–97
Presidencia da República do Brasil (PRB) (1965) Lei n° 4.711, de 15 de setembro de 1965: institui o novo código florestal. http://www.planalto.gov.br/ccivil_03/leis/L4771impressao.htm. Accessed 3 May 2017
Puga-Caballero A, MacGregor-Fors I, Ortega-Álvarez R (2014) Birds at the urban fringe: avian community shifts in different peri-urban ecotones of a megacity. Ecol Res 29:619–628
Reis E, Lopez-Iborra MG, Pinheiro RT (2012) Changes in bird species richness through different levels of urbanization: implications for biodiversity conservation and garden design in Central Brazil. Landsc Urban Plan 107:31–42
Reynaud P, Thioulouse J (2000) Identification of birds as biological markers along a neotropical urban–rural gradient (Cayenne, French Guiana), using co-inertia analysis. J Environ Manag 59:121–140
Rodríguez Tejerina M (2015) Sustainable cities in Latin America. IDDRI, working paper. Available via IDDRI. http://www.iddri.org/Publications/Collections/Idees-pour-le-debat/WP1615_EN.pdf. Accessed 3 May 2017
Rosselli L, Stiles FG (2012a) Local and landscape environmental factors are important for the conservation of endangered wetland birds in a high Andean plateau. Waterbirds 35:453–469
Rosselli L, Stiles FG (2012b) Wetland habitats of the Sabana de Bogotá Andean Highland Plateau and their birds. Aquat Conserv 22:303–317
Ruelas Inzunza E, Aguilar Rodríguez SH (2010) La avifauna urbana del Parque Ecológico Macuiltépetl en Xalapa, Veracruz, México. Ornitol Neotrop 21:87–103
Ruiz-Jaén M, Aide T (2006) An integrated approach for measuring urban forest restoration success. Urban For Urban Gree 4:55–68
Sanderson EW, Huron A (2011) Conservation in the city. Conserv Biol 25:421–423
Sandström UG, Angelstam P, Mikusiński G (2006) Ecological diversity of birds in relation to the structure of urban green space. Landsc Urban Plan 77:39–53
Sanz V, Caula S (2015) Assessing bird assemblages along an urban gradient in a Caribbean island (Margarita, Venezuela). Urban Ecosyst 18:729–746
Schneider A, Woodcock CE (2008) Compact, dispersed, fragmented, extensive? A comparison of urban growth in twenty-five global cities using remotely sensed data, pattern metrics and census information. Urban Stud 45:659–692
Schütz C, Schulze CH (2015) Functional diversity of urban bird communities: effects of landscape composition, green space area and vegetation cover. Ecol Evol 5:5230–5239

Seress G, Liker A (2015) Habitat urbanization and its effects on birds. Acta Zool Acad Sci Hung 61:373–408

Shanahan DF, Miller C, Possingham HP et al (2011) The influence of patch area and connectivity on avian communities in urban revegetation. Biol Conserv 144:722–729

Siemens AG (2010) Latin American green city index: assessing the environmental performance of Latin America's major cities. Available via Siemens. https://www.siemens.com/entry/cc/features/greencityindex_international/all/en/pdf/gci_report_summary.pdf. Accessed 3 May 2017

Silva CP, García CE, Estay SA et al (2015) Bird richness and abundance in response to urban form in a Latin American city: Valdivia, Chile as a case study. PLoS One 10:e0138120

Soga M, Yamamura Y, Koike S et al (2014) Land sharing vs. land sparing: does the compact city reconcile urban development and biodiversity conservation? J Appl Ecol 51:1378–1386

Sol D, Lapiedra O, González-Lagos C (2013) Behavioural adjustments for a life in the city. Anim Behav 85:1101–1112

Souza FL (2016) Booking.com®: a fake mega arboreal termitaria as an unusual nesting site for the Peach-fronted parakeet, *Eupsittula aurea* (Psittacidae). Biotemas 29:69–72

Sullivan BL, Aycrigg JL, Barry JH et al (2014) The eBird enterprise: an integrated approach to development and application of citizen science. Biol Conserv 169:31–40

Sullivan BL, Phillips T, Dayer A et al (2016) Using open access observational data for conservation action: a case study for birds. Biol Conserv 208:5–14

Toledo MCB, Donatelli RJ, Batista GT (2012) Relation between green spaces and bird community structure in an urban area in Southeast Brazil. Urban Ecosyst 15:111–131

Trujillo-Acosta A, Peraza-Estrella MJ, Marina-Hipolito JG et al (2017) Evaluación del corredor interurbano Río Torres, Costa Rica. Revista Forestal Mesoamericana Kurú 14:53–62

Turner WR, Nakamura T, Dinetti M (2004) Global urbanization and the separation of humans from nature. BioScience 54:585–590

Uras PMC, Gambera PMA, Pedro GK (2014) Avaliação do Manejo Arbóreo em dez parques Urbanos no Município de São Paulo. Paper presented at the XVIII Congressos Brasileiros de Arborização Urbana, Sociedade Brasileira de Arborização Urbana, Rio de Janeiro, 8–12 Nov 2014

Urban Management Programme for Latin America and the Caribbean (UMP-LAC) (2001) Towards participatory urban management in Latin American and Caribbean cities: a profile of the Urban Management Programme for Latin America and the Caribbean. Environ Urban 13:175–178

Villegas M, Garitano-Zavala Á (2010) Bird community responses to different urban conditions in La Paz, Bolivia. Urban Ecosyst 13:375–391

Wendel HEW, Zarger RK, Mihelcic JR (2012) Accessibility and usability: green space preferences, perceptions, and barriers in a rapidly urbanizing city in Latin America. Landsc Urban Plan 107:272–282

Ziller SR, Dechoum MS (2013) Plantas e vertebrados exóticos invasores em unidades de conservação no Brasil. Biodivers Bras 3:4–31

Chapter 9
Concluding Remarks: Current Knowledge and Future Directions

Ian MacGregor-Fors, Juan F. Escobar-Ibáñez, and Rafael Rueda-Hernández

Abstract In this concluding chapter, we underline the main conclusions of the reviewed topics regarding avian ecology in urban Latin America, namely: an updated literature review (Chap. 2); species richness and composition (Chap. 3); demography and population dynamics (Chap. 4); relationships with urban infrastructure (Chap. 5); behavior (Chap. 6); urban threats (Chap. 7); and conservation (Chap. 8). Afterward, we identify general knowledge gaps and future research directions, and conclude by highlighting some issues that require attention to enrich the succeeding study of bird ecology in urban Latin America. These issues include: (i) the consensus and use of standardized definitions for fundamental urban ecology concepts; (ii) considering broader temporal and spatial scale approaches; (iii) using updated analytical procedures; (iv) the need to strengthen our knowledge on urban bird distributions and natural history; and (v) taking a step forward in assessing the mechanistic processes behind the identified patterns. If we aim to create biodiverse, sustainable, and livable cities, we must reinforce the role of ecological knowledge in urban-related policies and strategies. Reinventing our cities will necessarily require willingness and proactive synergies among all the implied stakeholders, in which urban ecologists from all fields need to contribute with high-quality evidence-based information.

Keywords Biotic homogenization • Citywide surveys • Rural • Term standardization • Urban ecology

This book offers a general overview of avian urban ecology knowledge in Latin America, starting with an updated literature review (Chap. 2), following the one of Ortega-Álvarez and MacGregor-Fors (2011), afterward addressing some major topics, including species richness and composition (Chap. 3); demography and population dynamics (Chap. 4); relationships with urban infrastructure (mainly 'gray' and

I. MacGregor-Fors (✉) • J.F. Escobar-Ibáñez • R. Rueda-Hernández
Red de Ambiente y Sustentabilidad, Instituto de Ecología, A.C.
Carretera antigua a Coatepec 351, El Haya, Xalapa 91070, Veracruz, Mexico
e-mail: ian.macgregor@inecol.mx

I. MacGregor-Fors, J.F. Escobar-Ibáñez (eds.), *Avian Ecology in Latin American Cityscapes*, DOI 10.1007/978-3-319-63475-3_9

'green' components; Chap. 5); behavior (Chap. 6); urban threats (Chap. 7); and conservation (Chap. 8). Throughout the book, a representative sample of the available literature is reviewed and contextualized in relation to the findings from other regions of the globe, with each chapter highlighting the main research gaps for the topic and suggesting directions for further research. In this final chapter, we highlight the main conclusions of the reviewed topics (some of which have been stressed in the past; MacGregor-Fors and Ortega-Álvarez 2013), identify general common research gaps and directions, and underline some topic-specific needs related to highly relevant themes that require further attention, or in the worst of cases have passed unnoticed.

Considering the available literature (Chap. 2), together with the information provided in the subsequent chapters (Chaps. 3, 4, 5, 6, 7, and 8), it is evident that urban bird ecology in Latin America, although still underrepresented, is gaining momentum and recognition. Although there is a noticeable bias toward studies being carried out in populated countries (i.e., Brazil, Mexico, Argentina), studies from smaller countries are also emerging (e.g., Barbados, Guatemala, Bolivia). Most studies from these countries are focused on bird parasitism, reflecting the importance of this topic in these areas (see Chap. 7 for a thorough review and discussion). We also noticed a trend toward the study of exotic species, particularly House Sparrows (*Passer domesticus*), Rock Pigeons (*Columba livia*), European Starlings (*Sturnus vulgaris*), and Monk Parakeets (*Myiopsitta monachus*). Studies considered in this book highlight the relevance of focusing on these species in highly biodiverse regions, where their negative effects on native bird communities can be accentuated.

Fortunately, there is an increase in the number of ecological studies performed in urban Latin America in the last years; yet, species lists and new records have decreased in the last five years (Ortega-Álvarez and MacGregor-Fors 2011). Although this shortage is sadly not limited to urban birds studies in Latin America, applying to most ecology branches (Arnold 2003; Bury 2006), it is of crucial importance that Latin American ecologists intensify efforts to generate basic avian information in the region while continuing to study urban ecology patterns and related processes at different temporal and spatial scales. It is highly encouraging to find a growing number of topics addressed in the Latin American urban bird literature (see Chap. 2), including the role of noise in the establishment of tropical species in cities (e.g., Bermúdez-Cuamatzin et al. 2009; Ríos-Chelén et al. 2012), biotic homogenization related to urbanization (e.g., Filloy et al. 2015), multitaxonomic approaches to describe diversity patterns (e.g., MacGregor-Fors et al. 2016), as well as long-term studies (Vasconcelos et al. 2013). Given the nature of the Latin American region, its cities represent unique laboratories for studying birds, with our understanding of the way in which they respond to urbanization shedding important light in how cities should be managed and planned from an environmental perspective.

Specific research gaps and future directions focused on each topic reviewed in this book are highlighted in the corresponding chapters; however, there are some general research needs that have been consistently outlined in several chapters, reflecting the urgency to study general urban ecology patterns (for which information such as species lists and new records represent a baseline), and their related mechanistic processes, in broader temporal (i.e., throughout seasons, among years) and geographical scales (e.g., through urbanization intensity gradients, between countries, using citywide approaches; see Chaps. 2, 3, 4, 5, and 8). Another common remark is the call for more 'individual-to-community' approaches, where focal traits, such as avian behavior, population dynamics, and community shifts could strengthen our comprehension of the way in which birds respond to urbanization in Latin America. These studies would then help to clarify similarities and differences when contrasted with patterns from other regions of the world.

Not surprisingly, most of the publications focused on urban birds in Latin America are from some of the most populated countries of the region that have long been urban (e.g., Mexico, Brazil, Argentina; Montgomery 2008; United Nations 2015). It is noticeable that there is a general underrepresentation of countries from Central America and the Caribbean. Even so, in those countries, where studies have been carried out, there is an important geographical bias related to focal cities and regions, such as southeastern Brazil, central Mexico (i.e., Trans-Mexican Volcanic Axis), and the Metropolitan Region of Buenos Aires, Argentina. Although knowledge from large and focal cities is crucial in understanding how wildlife responds to urbanization, the lack of information on the ecology of small-to-medium-sized urban settlements is a critical knowledge gap if our goal is to merge the ecological sphere to the physical and social ones in which urban development and sprawling are currently happening (Berkowitz et al. 2003; McDonnell and MacGregor-Fors 2016). While urban areas may seem similar in structure (Batty 2008), their origin, development, and dynamics make each of them quite unique (Davis 1955; Pascal 1987; Oke 2006). Thus, to understand such complex phenomena, we need to address ecological questions in cities of different size, origin, age, latitude, and altitude, with those located in highly biodiverse and understudied regions being of critical importance.

Although urban bird ecology in Latin America has an important lag when compared with the knowledge pool of developed temperate regions (~30 years sensu Ortega-Álvarez and MacGregor-Fors 2011), the current status is encouraging and is already underpinning interesting general ecological patterns including cities at different latitudes. When comparing results of studies performed in urban Latin America with those from other regions of the world, some patterns hold (see Chap. 2), but some important exceptions have also been identified, such as high species richness values in greenspaces located near and far from the city center, with those located at intermediate distances showing lower values (e.g., Faggi and Perepelizin 2006); large number of species reported within city limits (e.g., 341 species in Xalapa, Mexico: González-García 2014, 2016; 203 species in Pereira, Colombia: Londoño-Betancourth 2011; 186 species in Santa Marta, Colombia: Strewe et al.

2009), and important representation of insectivores and frugivores, including large species such as toucans and macaws, as part of the urban bird communities.

Besides what has been already said above and in the rest of the book, we have also combined pieces of information and advice from the chapters to identify some additional aspects that are important to consider as they could aid in the consolidation of urban bird ecology in Latin America (with some of them applying globally). The first issue deals with the standardization of urban-related fundamental concepts. Despite the effort of some urban ecologists throughout the globe to clearly define basic concepts, the lack of use of standard definitions in the literature prevents us from fully understanding, for instance, the nature of study sites. That is the case of the 'urban' concept, which has been approached in the past from the urban studies trench (e.g., McIntyre et al. 2000; Marzluff et al. 2001; MacGregor-Fors 2011); yet, the use of this concept based on the governmental definitions (see United Nations 2005 for a detailed global analysis) has importantly permeated the ecological literature. There are also a considerable number of terms related to urban locations and/or conditions (e.g., peri-urban, intra-urban, extra-urban, exurban, urban core), as well as land-uses throughout urbanized landscapes, that lack standard definitions, yet are often used to describe the context of study areas, for example, urban–rural gradients (see McDonnell and Hahs 2008 for a thorough review). The latter leads us to another crucial concept in the literature: the widely varying conditions related to the 'rural' land-use. This term has been used to describe study sites from a wide array of environmental conditions, ranging from native vegetation (tropical forest through xerophytic vegetation) to human-modified systems (e.g., grasslands, agricultural fields) (Teta et al. 2012; Calegaro-Marques and Amato 2014; Rodríguez-Martínez et al. 2014; Sanz and Caula 2015). Given that most studies using the gradient approach measure 'built cover' as the gradually changing variable, we suggest they may be better described as 'urbanization intensity' or 'built cover' gradients, always specifying the nature of the extreme with low-to-null urbanized surface. If the gradually changing variable is not a measure or proxy of 'built cover', but relates instead to vegetation, it is crucial to specify which is the gradually changing variable and to thoroughly describe the context of the study area given that systems considered as 'rural' can be highly contrasting. If the comparisons are not related to a gradually changing variable, but rather categorical comparisons, such as use-based categories (e.g., urban–agricultural contrast, commercial–residential–greenspace comparison), then it is imperative to state them as such and avoid the use of a gradient framework. We recognize that this lack of standard definitions for terms is not unique to the Latin American region. Indeed, if we aim to create a pool of knowledge that allows generalizations and the identification of particularities that enrich our contributions to evidence-based knowledge in urban management and planning, we need to add to the global discussion and stress the uniqueness of Latin American urban systems. Doing this effectively requires working with a broader, more realistic, vision of urban ecosystems and its fundamental concepts.

The second topic we would like to stress has two closely related parts: experimental design and analytical procedures. The experimental design for many studies has focused on assessing a specific question in a particular greenspace or residential area, or has followed urban–rural gradient approaches. These studies have been conducted under the conceptual framework of 'ecology in the city', answering ecological questions within, or considering, urban systems (Pickett et al. 2016). Undoubtedly, such information is of great value in adding to our comprehension of the response of birds to urbanization in Latin America. Nevertheless, it is imperative to move toward an 'ecology of the city' approach, seeking to understand the complex interactions between the ecological, physical, and social spheres of urban systems (Berkowitz et al. 2003; Pickett et al. 2016). Although additional research on urban birds can only contribute to strengthen our knowledge on the ecological sphere of urban systems, it is essential to have a fair representation of cities, for which citywide surveys have been proposed. These types of surveys allow the inclusion of a representative sample of the spatial heterogeneity of the ecological, infrastructure, and social condition of cities (Falfán and MacGregor-Fors 2016). As far as we know, only five urban bird studies performed in urban Latin America have followed this approach (Fontana et al. 2011; MacGregor-Fors et al. 2012; Sacco et al. 2015; Silva et al. 2015; Escobar-Ibáñez and MacGregor-Fors 2016). Results of Silva et al. (2015) highlight the importance of several greenspace types for native bird species richness, finding a positive relationship throughout the city with non-municipal greenspaces, but a surprising negative relationship with private gardens. Also, Fontana et al. (2011) show that human population density, a variable that has been often used as a proxy of urbanization intensity (as well as in many other publications; e.g., McDonnell and Hahs 2008; Strohbach et al. 2009), explains less variation in bird species richness and structure than the tree density and urban noise did, showing the importance of citywide surveys in understanding the response of birds to urbanization. Regarding procedures to analyze urban bird ecology data, it is crucial that researchers take the leap in embracing robust and updated methods, particularly those related to avian diversity (see Chap. 3 for suggestions regarding methodological approaches).

Third, we would like to summarize and highlight the importance of conducting urban bird ecology studies that address patterns at different spatial and temporal scales. Even though research from urban Latin America is entering the larger global knowledge pool and helping to shape a broader understanding of urban ecology, considering multicity, regional-to-global approaches, taking seasonality into account, as well as a wider array of explanatory variables (e.g., food availability, noise, light pollution, species interactions), untangling the mechanistic processes behind the patterns is essential to advance our understanding of urban avian ecology (Shochat et al. 2006; McDonnell and Hahs 2013). Although Shochat et al. (2006) emphasized this urgent need a decade ago, large-scale comparative studies, as well as those unraveling the drivers behind bird responses to urbanization, have been slow, not just in Latin America, but also worldwide. Thus, in addition to the relative scarcity of basic ornithology publications (e.g., natural history, species lists, new records) from urban Latin America, urban bird ecologists in the region ought to intensify their efforts, describing basic natural history traits and outlining species lists from poorly known areas, while performing ecological studies seeking to describe regional urban bird ecological patterns and identify the related processes.

Finally, we would like to bring attention to a topic that has a particularly urgent relevance for this region due to its pace of urbanization, increasing human population, economic inequality, and outstanding bird diversity (Stotz et al. 1996; Santandreu et al. 2002; United Nations 2015): biotic homogenization related to urbanization. Although studies have stated that urbanization is a major cause of biotic homogenization (e.g., McKinney 2006), recent global studies show contrasting results. On the one hand, Aronson et al. (2014) show that despite urbanization has a drastic effect on bird diversity and several exotic bird species are shared throughout cities around the globe, when comparing incidence-based data, there is no evidence that taxonomic homogenization has occurred at the global scale. Opposedly Sol et al. (2014) suggest that, based on their results, species filtering caused by urbanization reduces beta-diversity and promotes biotic homogenization. Studies performed at regional and intracity levels, including Latin America, have found higher taxonomic similarity between urban land-uses when compared with greenspaces, as well as increased functional similarity when comparing urban and nonurban locations (e.g., Ortega-Álvarez and MacGregor-Fors 2009). Yet, it seems that some studies are not assessing the homogenization process, but are instead evaluating species-filtering effects of urbanization. As defined by Olden and Rooney (2006), biotic homogenization "refers to the process by which the genetic, taxonomic or functional similarities of regional biotas increase over time" (p. 133). Although it is well known that urbanization has a huge impact on bird species richness (Chace and Walsh 2006; Evans et al. 2009; Aronson et al. 2014; Sol et al. 2014), species filtering through a 'semi-permeable' process of birds not being able to cross the urban–nonurban ecotone (MacGregor-Fors 2010) does not necessarily imply that urban biotas are more similar among them and that they do not include a representative sample of the regional species pool. Regarding a functional approach, we suggest to follow fine and descriptive schemes to determine avian functionality, such as diet types following Lopes et al. (2016). One important thing to consider is whether biotic homogenization should be assessed only through incidence-based comparisons, or if it is valid to use an abundance-based approach. From our perspective, using abundance-based analyses to test bird taxonomic and/or functional homogenization in urban areas could be misleading, as the numbers of a few urban dwellers, often exotic and/or invasive species that experience important population growth in cities and often comprise important proportion of total urban bird numbers, could considerably influence the results. We recognize that biotic homogenization can occur when the abundance or functional attributes are dominated by a few species, also pinpointed by Olden and Rooney (2006). Nevertheless, in doing so, we could be failing to detect the importance and uniqueness of outstandingly bird-rich cities with different avifaunas when compared with other cities in the regional and national scale. That is the case of the city of Xalapa, Mexico, with records for 341 bird species within the city borders in the last three decades, once the latest records have been taken into account (González-García et al. 2016). Regardless of the huge bird diversity recorded in Xalapa, the overall bird community is biased by the important representation of two species, an exotic invader (House Sparrow) and a native generalist (Great-tailed Grackle—*Quiscalus mexicanus*), comprising ~42% of the total recorded abundance in a citywide survey (Escobar-Ibáñez and MacGregor-Fors 2016). Using abundance-based approaches would result in high similarities with nearby (e.g., Veracruz, Puebla) and farther cities (e.g., Mexico City, Guadalajara, Morelia), where the sparrow and grackle are abundant; yet, Xalapa's avifauna is quite unique, even on a national scale. It is noteworthy that we are, by no means, suggesting that urbanization cannot be a source of avian taxonomic and/or functional homogenization. Rather, we consider that studies assessing this important ecological pattern and its related processes should be cautious, considering all the details, and including comparative frameworks, required to appropriately tackle the question.

The information provided in this book sets readers face-to-face with the current state of knowledge for bird studies in urban Latin America, highlighting the need to use common schemes and languages based on updated methodologies to fill in the many remaining knowledge gaps and consolidate our advancement in this field. If our aim is to create biodiverse livable cities, we must integrate ecological knowledge into urban management and planning policies and strategies (Niemelä 1999; McDonnell and MacGregor-Fors 2016). For this, as urban ecology research grows, there is an urgent need to communicate our findings to policy-makers, land managers, planners, and the general public to deliver stronger biodiversity and ecology outcomes associated with urban development. But for evidence-based knowledge to permeate and transform into action, willingness of all implied stakeholders is essential (MacGregor-Fors 2015). We urge Latin American urban ecologists, as well as experts in other urban-related fields, to think out of the box. We not only need to lay our cards on the table in the form of scientific-based conservation strategies and urban development regulations that focus not only in greenspaces, but also consider streetscapes and the rest of the urban system, playing a stronger role in contributing to the decision-making processes that are shaping our cities. This will not only promote bird-friendly cities, but will also increase the environmental quality of cities, as well as providing improved outcomes for human health and well-being (Miller 2005; MacGregor-Fors et al. 2009).

Acknowledgments The authors are most grateful to Amy K. Hahs for her comments and suggestions of an earlier version of this chapter, which greatly enhanced its quality and clarity, as well as Mark J. McDonnell for enlightening discussions regarding some of the developed ideas. JFE-I and RR-H acknowledge the scholarship and financial support provided by the National Council of Science and Technology (CONACYT 366146, 171086, respectively), as well as the Graduate School of the Instituto de Ecología, A.C. (INECOL).

References

Arnold SJ (2003) Too much natural history, or too little? Anim Behav 65:1065–1068

Aronson MFJ, La Sorte FA, Nilon CH et al (2014) A global analysis of the impacts of urbanization on bird and plant diversity reveals key anthropogenic drivers. Proc R Soc Lond B Biol Sci 281:20133330

Batty M (2008) The size, scale, and shape of cities. Science 319:769–771

Berkowitz AR, Nilon CH, Hollweg KS (eds) (2003) Understanding urban ecosystems: a new frontier for science and education. Springer, New York

Bermúdez-Cuamatzin E, Ríos-Chelén AA, Gil D et al (2009) Strategies of song adaptation to urban noise in the House Finch: syllable pitch plasticity or differential syllable use? Behaviour 146:1269–1286

Bury BR (2006) Natural history, field ecology, conservation biology and wildlife management: time to connect the dots. Herpetol Conserv Biol 1:56–61

Calegaro-Marques C, Amato SB (2014) Urbanization breaks up host-parasite interactions: a case study on parasite community ecology of Rufous-bellied Thrushes (*Turdus rufiventris*) along a rural-urban gradient. PLoS One 9:e103144

Chace JF, Walsh JJ (2006) Urban effects on native avifauna: a review. Landsc Urban Plan 74:46–69

Davis K (1955) The origin and growth of urbanization in the world. Am J Sociol 60:429–437

Escobar-Ibáñez JF, MacGregor-Fors I (2016) Peeking in to the past to plan the future: assessing bird species richness in a neotropical city. Urban Ecosyst 19:657–667

Evans KL, Newson SE, Gaston KJ (2009) Habitat influences on urban avian assemblages. Ibis 151:19–39

Faggi A, Perepelizin P (2006) Bird richness along an urbanization gradient in Buenos Aires city. Rev Mus Argentino Cien Nat ns 8:289–297

Falfán I, MacGregor-Fors I (2016) Woody neotropical streetscapes: a case study of tree and shrub species richness and composition in Xalapa. Madera Bosques 22:95–110

Filloy J, Grosso S, Bellocq MI (2015) Urbanization altered latitudinal patterns of bird diversity-environment relationships in the southern Neotropics. Urban Ecosyst 18:777–791

Fontana CS, Burger MI, Magnusson WE (2011) Bird diversity in a subtropical South-American city: effects of noise levels, arborisation and human population density. Urban Ecosyst 14:341–360

González-García F, Straub R, Lobato-García JA et al (2014) Birds of a neotropical green city: an up-to-date review of the avifauna of the city of Xalapa with additional unpublished records. Urban Ecosyst 17:991–1012

González-García F, Straub R, Lobato-García JA et al (2016) Nuevos registros y notas adicionales comentadas sobre la avifauna de la ciudad de Xalapa, Veracruz, México. Acta Zool Mex ns 32:253–269

Londoño-Betancourth JC (2011) Una mirada a la diversidad ornitológica de Pereira. Bol Cient Mus Hist Nat 15:84–103

Lopes LE, Fernandes AM, Medeiros MCI et al (2016) A classification scheme for avian diet types. J Field Ornithol 87:309–322

MacGregor-Fors I (2010) How to measure the urban-wildland ecotone: redefining 'peri-urban' areas. Ecol Res 25:883–887

MacGregor-Fors I (2011) Misconceptions or misunderstandings? On the standardization of basic terms and definitions in urban ecology. Landsc Urban Plan 100:347–349

MacGregor-Fors I (2015) 10 scientists and 10 practitioners walk into a bar...what would they talk about? How can research and knowledge generation be co-created to better support practitioners and evidence-based decision making? In: Maddox D (ed) Global roundtable. The nature of cities. http://www.thenatureofcities.com/2015/06/30/10-scientists-and-10-practitioners-walk-into-a-barwhat-would-they-talk-about-how-can-research-and-knowledge-generation-be-co-created-to-better-support-practitioners-and-evidence-based-decisi/. Accesed 24 Jan 2017

MacGregor-Fors I, Ortega-Álvarez R (eds) (2013) Ecología urbana: experiencias en América Latina. Available via INECOL. http://www1.inecol.edu.mx/libro_ecologia_urbana/. Accessed 19 Jan 2017

MacGregor-Fors I, Ortega-Álvarez R, Schondube JE (2009) On the ecological quality of urban systems: an ornithological perspective. In: Graber DS, Birmingham KA (eds) Urban planning in the 21st century. Nova Science Publishing, New York, pp 51–66

MacGregor-Fors I, Morales-Pérez L, Schondube J (2012) From forests to cities: effects of urbanization on tropical birds. In: Lepzcyk CA, Warren PS (eds) Urban bird ecology and conservation. University of California Press, Berkeley, pp 33–48

MacGregor-Fors I, Escobar F, Rueda-Hernández R et al (2016) City "green" contributions: the role of urban greenspaces as reservoirs for biodiversity. Forests 7:146
Marzluff JM, Bowman R, Donnelly R (2001) A historical perspective on urban bird research: trends, terms, and approaches. In: Marzluff JM, Bowman R, Donnelly R (eds) Avian ecology and conservation in an urbanizing world. Springer, New York, pp 1–17
McDonnell MJ, Hahs AK (2008) The use of gradient analysis studies in advancing our understanding of the ecology of urbanizing landscapes: current status and future directions. Landsc Ecol 23:1143–1155
McDonnell MJ, Hahs AK (2013) The future of urban biodiverstity research: moving beyond the 'low-hanging fruit'. Urban Ecosyst 16:397–409
McDonnell MJ, MacGregor-Fors I (2016) The ecological future of cities. Science 352:936–938
McIntyre NE, Knowles-Yánez K, Hope D (2000) Urban ecology as an interdisciplinary field: differences in the use of "urban" between the social and natural sciences. Urban Ecosyst 4:5–24
McKinney ML (2006) Urbanization as a major cause of biotic homogenization. Biol Conserv 127:247–260
Miller JR (2005) Biodiversity conservation and the extinction of experience. Trends Ecol Evol 20:430–434
Montgomery MR (2008) The urban transformation of the developing world. Science 319:761–764
Niemelä J (1999) Ecology and urban planning. Biodivers Conserv 8:119–131
Oke TR (2006) Towards better scientific communication in urban climate. Theor Appl Climatol 84:179–190
Olden JD, Rooney TP (2006) On defining and quantifying biotic homogenization. Glob Ecol Biogeogr 15:113–120
Ortega-Álvarez R, MacGregor-Fors I (2009) Living in the big city: effects of urban land-use on bird community structure, diversity, and composition. Landsc Urban Plan 90:189–195
Ortega-Álvarez R, MacGregor-Fors I (2011) Dusting-off the file: a review of knowledge on urban ornithology in Latin America. Landsc Urban Plan 101:1–10
Pascal A (1987) The vanishing city. Urban Stud 24:597–603
Pickett STA, Cadenasso ML, Childers DL et al (2016) Evolution and future of urban ecological science: ecology *in*, *of*, and *for* the city. Ecosyst Health Sustain 2:e01229
Ríos-Chelén AA, Salaberria C, Barbosa I et al (2012) The learning advantage: bird species that learn their song show a tighter adjustment of song to noisy environments than those that do not learn. J Evol Biol 25:2171–2180
Rodriguez Martínez S, Carrete M, Roques S et al (2014) High urban breeding densities do not disrupt genetic monogamy in a bird species. PLoS One 9:e91314
Sacco AG, Rui AM, Bergmann FB et al (2015) Perda de diversidade taxonômica e funcional de aves em area urbana no sul do Brasil. Iheringia Ser Zool 105:276–287
Santandreu A, Gómez Perazzoli A, Dubbeling M (2002) Biodiversity, poverty and urban agriculture in Latin America. UA-Mag 6
Sanz V, Caula S (2015) Assessing bird assemblages along an urban gradient in a Caribbean island (Margarita, Venezuela). Urban Ecosyst 18:729–746
Shochat E, Warren PS, Faeth SH et al (2006) From patterns to emerging processes in mechanistic urban ecology. Trends Ecol Evol 21:186–191
Silva CP, García CE, Estay SA et al (2015) Bird richness and abundance in response to urban form in a Latin American city: Valdivia, Chile as a case study. PLoS One 10:e0138120
Sol D, González-Lagos C, Moreira D et al (2014) Urbanisation tolerance and the loss of avian diversity. Ecol Lett 17:942–950
Stotz DF, Fitzpatrick JW, Parker TA et al (eds) (1996) Neotropical birds: ecology and conservation. University of Chicago Press, Chicago
Strewe R, Villa de León C, Alzate J (2009) Las aves del Campus de la Universidad del Magdalena, Santa Marta, Colombia. Intropica 4:79–91
Strohbach M, Haase D, Kabisch N (2009) Birds and the city: urban biodiversity, land use, and socioeconomics. Ecol Soc 14:31

Teta P, Hercolini C, Cueto G (2012) Variation in the diet of Western Barn Owls (*Tyto alba*) along an urban-rural gradient. Wilson J Ornithol 124:589–596

United Nations (2005) Demographic yearbook. Statistics Division of the United Nations Department of Economic and Social Affairs, New York

United Nations (2015) World population 2015. Population Division Department of Economic and Social Affairs United Nations, New York

Vasconcelos MF, de Carvalho E, Mazzoni LG et al (2013) Long-term avifaunal survey in an urban ecosystem from Southeastern Brazil, with comments on range extensions, new and disappearing species. Pap Avulsos de Zool (São Paulo) 53:327–344

Index

I. MacGregor-Fors, J.F. Escobar-Ibáñez (eds.), *Avian Ecology in Latin American Cityscapes*, DOI 10.1007/978-3-319-63475-3

C

D

E

F

G

V

W

MIX
Papier aus verantwortungsvollen Quellen
Paper from responsible sources
FSC® C105338

If you have any concerns about our products,
you can contact us on
ProductSafety@springernature.com

In case Publisher is established outside the EU,
the EU authorized representative is:
Springer Nature Customer Service Center GmbH
Europaplatz 3, 69115 Heidelberg, Germany

Printed by Libri Plureos GmbH
in Hamburg, Germany